Triumph Tiger Cub and Terrier Owners Workshop Manual

by Pete Shoemark

Models covered:

T15 Terrier 149 cc 1952 - 1956
T20 Tiger Cub 199 cc 1954 - 1966
T20C Tiger Cub Competition 199 cc 1956 - 1959
T20S Tiger Cub Sports 199 cc 1959 - 1960
T20T Tiger Cub Trials 199 cc 1960 - 1961
T20SL Tiger Cub Scrambler 199 cc 1960 - 1961
T20SS Tiger Cub Sports 199 cc 1961 - 1963
T20SH Tiger Cub Sports 199 cc 1961 - 1966
TR20 Tiger Cub Trials 199 cc 1963 - 1965
T20SC Super Cub/Bantam Cub 199 cc 1966 - 1968

ISBN 978 0 85696 414 5

(414-9L1)

Haynes Group Limited
Haynes North America, Inc

www.haynes.com

Acknowledgements

Grateful thanks are due to Mr P. Warren, who provided the machine used for the photographic sequence, and to Johnnie Edwards of Richmond, whose machine is featured on the front cover, for his enthusiastic assistance with the photographic session.

Eric Raymond of Templecombe loaned most of the technical literature used in the compilation of the manual. The new gearbox sprocket was supplied by Supersprox, Knighton, Radnorshire, and various spares were supplied by A Bennett and Son, Station Street Garage, Atherstone, Warwicks.

Martin Penny assisted with the stripdown and rebuilding sequences, and devised various ingenious methods for overcoming the lack of factory service tools. Les Brazier arranged and took the photographs. Jeff Clew edited the text.

Our thanks are also due to the Avon Rubber Company, who supplied information on tyre fitting, to NGK Spark Plugs (UK) Limited for advice about sparking plug conditions and Renold Ltd for advice on chain care and renewal. Mr G Binks, of Amal Ltd, provided specification information on Amal carburettors, which would have otherwise proved unobtainable.

About this manual

The author of this manual has the conviction that the only way in which a meaningful and easy to follow text can be written is first to carry out the work himself, under conditions that prevail in the normal average household. As a result, the hands seen in the photographs are those of the author.

The machine used for the photographic sequences was not new, but had covered several thousand miles. The reason for this was so that the conditions encountered would be the same as those seen by the average user. Also, unless specially mentioned, and therefore considered essential, Triumph service tools have not been used. There is invariably an alternative means of removing or slackening a component when service tools are not available and the risk of damage has to be avoided at all costs.

This manual is divided into numbered sections, in each Chapter. Cross reference throughout this manual is quite straightforward and logical. When reference is made, 'See Section 6.10', it means Section 6, paragraph 10, in the same Chapter. If another Chapter were meant it would say 'See Chapter 2, Section 6:10'.

All the photographs are captioned with a number which denotes the section paragraph number to which they refer, and are always relevant to the Chapter text adjacent. Figure numbers (usually line illustrations) appear in numerical order within a given Chapter. 'Fig. 1.1' therefore refers to the first figure in Chapter 1.

Left-hand and right-hand descriptions of the machine and components refer to the left and right of the machine with the rider normally seated.

Motorcycle manufacturers continually make changes to specifications and recommendations, and these, when notified, are incorporated into our manuals at the earliest opportunity.

Whilst every care is taken to ensure that the information in this manual is correct no liability can be accepted by the authors or publishers for loss, damage or injury caused by any errors in or omissions from the information given.

Contents

Chapter	Section	Page	Section	Page
Introductory sections	Acknowledgements	2	Ordering spare parts	8
	About this manual	2	Routine maintenance	9
	Safety first!	5	Recommended lubricants	
	Introduction to the Triumph		and adjustment settings	13
	Tiger Cub range	6	Working conditions and tools	14
	Dimensions and weights	6		
Chapter 1 Engine, clutch and gearbox	Operations: Unit in frame	17	Examination and renovation:	
	Engine removal	17	Rockers, Spindles and covers	36
	Engine removal	17	Gearbox and kickstart	36
	Dismantling – general	19	Clutch	38
	Examination and renovation	30	Engine reassembly – general	40
	Decarbonisation	34–35	Fitting unit in frame	49
	Valve grinding	34	Running a rebuilt unit	50
Chapter 2 Fuel system and lubrication	Petrol tank	55	Air filter	63
	Petrol tap	55	Exhaust system	63
	Carburettor removal	55	Oil pump	64
	Carburettor settings	63	Rocker oil feed	65
Chapter 3 Ignition system	Energy-transfer ignition	69	Distributor	71
	Contact breaker	69	Ignition timing	71
	Automatic timing unit	69	Ignition coil – checking	73
	Condenser	71	Sparking plugs	74
Chapter 4 Frame and forks	Front forks legs	75	Plunger rear suspension	82
	Steering head assembly	76	Swinging arm rear suspension	82
	Fork forks	78	Rear suspension unit	85
	Steering head bearings	81	Centre stand	85
	Frame assembly	82	Steering head lock	86
Chapter 5 Wheels, brakes and tyres	Front wheel	89	Rear wheel bearings	92
	Front brake	89	Rear wheel sprocket	94
	Wheel bearings	89	Final drive chain	94
	Rear wheel	91	Wheel balance	94
	Rear brake	91	Tyres	94 – 96
Chapter 6 Electrical system	Checking the charging system	98	Bulbs – replacement	101–102
	Rectifier	100	Horn – adjustment	102
	Alternator– location	100	Wiring – layout	103
	Battery – charging	101	Headlamp and Ignition switches	103
	Headlamp	101	Wiring diagrams	104–106

Note: General descriptions and specifications are given in each Chapter immediately after list of contents.
Fault diagnosis is given when applicable at the end of the Chapter.

Conversion factors   107
English/American terminology   108
Index   109 – 111

1963 T20 Triumph Tiger Cub

1962 T20SH Triumph Tiger Cub

Safety First!

Professional motor mechanics are trained in safe working procedures. However enthusiastic you may be about getting on with the job in hand, do take the time to ensure that your safety is not put at risk. A moment's lack of attention can result in an accident, as can failure to observe certain elementary precautions.

There will always be new ways of having accidents, and the following points do not pretend to be a comprehensive list of all dangers; they are intended rather to make you aware of the risks and to encourage a safety-conscious approach to all work you carry out on your vehicle.

Essential DOs and DON'Ts

DON'T start the engine without first ascertaining that the transmission is in neutral.

DON'T suddenly remove the filler cap from a hot cooling system — cover it with a cloth and release the pressure gradually first, or you may get scalded by escaping coolant.

DON'T attempt to drain oil until you are sure it has cooled sufficiently to avoid scalding you.

DON'T grasp any part of the engine, exhaust or silencer without first ascertaining that it is sufficiently cool to avoid burning you.

DON'T allow brake fluid or antifreeze to contact the machine's paintwork or plastic components.

DON'T syphon toxic liquids such as fuel, brake fluid or antifreeze by mouth, or allow them to remain on your skin.

DON'T inhale dust — it may be injurious to health (see *Asbestos* heading).

DON'T allow any spilt oil or grease to remain on the floor — wipe it up straight away, before someone slips on it.

DON'T use ill-fitting spanners or other tools which may slip and cause injury.

DON'T attempt to lift a heavy component which may be beyond your capability — get assistance.

DON'T rush to finish a job, or take unverified short cuts.

DON'T allow children or animals in or around an unattended vehicle.

DON'T inflate a tyre to a pressure above the recommended maximum. Apart from overstressing the carcase and wheel rim, in extreme cases the tyre may blow off forcibly.

DO ensure that the machine is supported securely at all times. This is especially important when the machine is blocked up to aid wheel or fork removal.

DO take care when attempting to slacken a stubborn nut or bolt. It is generally better to pull on a spanner, rather than push, so that if slippage occurs you fall away from the machine rather than on to it.

DO wear eye protection when using power tools such as drill, sander, bench grinder etc.

DO use a barrier cream on your hands prior to undertaking dirty jobs — it will protect your skin from infection as well as making the dirt easier to remove afterwards; but make sure your hands aren't left slippery.

DO keep loose clothing (cuffs, tie etc) and long hair well out of the way of moving mechanical parts.

DO remove rings, wristwatch etc, before working on the vehicle — especially the electrical system.

DO keep your work area tidy — it is only too easy to fall over articles left lying around.

DO exercise caution when compressing springs for removal or installation. Ensure that the tension is applied and released in a controlled manner, using suitable tools which preclude the possibility of the spring escaping violently.

DO ensure that any lifting tackle used has a safe working load rating adequate for the job.

DO get someone to check periodically that all is well, when working alone on the vehicle.

DO carry out work in a logical sequence and check that everything is correctly assembled and tightened afterwards.

DO remember that your vehicle's safety affects that of yourself and others. If in doubt on any point, get specialist advice.

IF, in spite of following these precautions, you are unfortunate enough to injure yourself, seek medical attention as soon as possible.

Asbestos

Certain friction, insulating, sealing, and other products — such as brake linings, clutch linings, gaskets, etc — contain asbestos. *Extreme care must be taken to avoid inhalation of dust from such products since it is hazardous to health.* If in doubt, assume that they *do* contain asbestos.

Fire

Remember at all times that petrol (gasoline) is highly flammable. Never smoke, or have any kind of naked flame around, when working on the vehicle. But the risk does not end there — a spark caused by an electrical short-circuit, by two metal surfaces contacting each other, or even by static electricity built up in your body under certain conditions, can ignite petrol vapour, which in a confined space is highly explosive.

Always disconnect the battery earth (ground) terminal before working on any part of the fuel or electrical system, and never risk spilling fuel on to a hot engine or exhaust.

It is recommended that a fire extinguisher of a type suitable for fuel and electrical fires is kept handy in the garage or workplace at all times. Never try to extinguish a fuel or electrical fire with water.

Fumes

Certain fumes are highly toxic and can quickly cause unconsciousness and even death if inhaled to any extent. Petrol (gasoline) vapour comes into this category, as do the vapours from certain solvents such as trichloroethylene. Any draining or pouring of such volatile fluids should be done in a well ventilated area.

When using cleaning fluids and solvents, read the instructions carefully. Never use materials from unmarked containers — they may give off poisonous vapours.

Never run the engine of a motor vehicle in an enclosed space such as a garage. Exhaust fumes contain carbon monoxide which is extremely poisonous; if you need to run the engine, always do so in the open air or at least have the rear of the vehicle outside the workplace.

The battery

Never cause a spark, or allow a naked light, near the vehicle's battery. It will normally be giving off a certain amount of hydrogen gas, which is highly explosive.

Always disconnect the battery earth (ground) terminal before working on the fuel or electrical systems.

If possible, loosen the filler plugs or cover when charging the battery from an external source. Do not charge at an excessive rate or the battery may burst.

Take care when topping up and when carrying the battery. The acid electrolyte, even when diluted, is very corrosive and should not be allowed to contact the eyes or skin.

If you ever need to prepare electrolyte yourself, always add the acid slowly to the water, and never the other way round. Protect against splashes by wearing rubber gloves and goggles.

Mains electricity

When using an electric power tool, inspection light etc which works from the mains, always ensure that the appliance is correctly connected to its plug and that, where necessary, it is properly earthed (grounded). Do not use such appliances in damp conditions and, again, beware of creating a spark or applying excessive heat in the vicinity of fuel or fuel vapour.

Ignition HT voltage

A severe electric shock can result from touching certain parts of the ignition system, such as the HT leads, when the engine is running or being cranked, particularly if components are damp or the insulation is defective. Where an electronic ignition system is fitted, the HT voltage is much higher and could prove fatal.

Introduction to the Triumph Tiger Cub range

The story of the Tiger Cub can be traced back to early in 1952, when Triumph announced the introduction of a 150 cc ohv lightweight machine called the Terrier. This machine proved to be something of a surprise, as it was aimed at a market traditionally dominated by small, and often rather crude, two-stroke machines. By the time that the Earls Court Show took place that year, it was obvious that the Terrier was destined for success.

The pushrod operated overhead valve engine had a bore and stroke of 57 mm x 58.5 mm, which was relatively unusual in an era when under-square, or long stroke, engines were the general rule. This configuration made for an engine unit which was both light in weight and a brisk performer. Edward Turner had intended to achieve this when he set out to equal the performance of the pre-war 250 cc single, with a smaller, lighter and more economical machine. In production form, the engine produced 8.3 bhp at 6,500 rpm. More importantly, a flat spread of torque was produced, ranging from 150 lb f in at 2,750 rpm, to 185 lb f in at 5,000 rpm. Thus the Terrier was very flexible, perhaps its main advantage over its two-stroke competitors.

The 199 cc T20 Tiger Cub first appeared in 1954, and was similar to the Terrier, apart from its increased capacity. Bore and stroke were increased to 63 mm x 64 mm, maintaining the almost square configuration of the original engine.

During the fourteen years of the Tiger Cub's production, numerous minor and major changes took place, and various competition and sporting versions of the basic T20 roadster were introduced.. By 1968, the Tiger Cub had reached its final form, and by then sported a BSA Bantam frame. In October of that year, production of the model finally ceased. Its passing was noted with some sadness by the many motorcyclists who had owned one of the once ubiquitous Cubs.

Despite its being obsolete for ten years, at the time of writing, there are still large numbers of these machines alive and well in the UK and overseas. This manual is intended to help owners to keep their machines that way. Although aimed in general at the Tiger Cub range, it will be appreciated that the Terrier models were very similar to the early Tiger Cubs, and where possible, relevant information has been included in the Specifications and in the main text.

It should be noted that various performance conversion parts were available from the manufacturers and from other sources, and care must be taken to ensure that the correct valve and ignition timing settings are used to suit the various cams, pistons and cylinder heads used. It follows, then, that the permutations are almost infinite, and whilst as much information as is available is included, it is recommended that replacement parts, especially secondhand items, are carefully compared with a known standard component, to ensure suitability.

Dimensions and weights

Seat height	28½ – 30 in (720 – 762 mm)
Wheelbase	49 in (1245 mm)
Length	77 – 78·5 in (1960 – 1990 mm)
Width	25 – 26 in (635 – 660 mm)
Overall height	36 – 38 in (910 – 965 mm)
Ground clearance	4·5 – 6 in (114 – 152 mm)
Weight	205 – 223 lbs (92 – 101 kg)

Note: The above figures are given as a rough guide, and may vary between the values shown depending on the model.

Modifications to the
Triumph Terrier and Tiger Cub range

T15 Terrier

Introduced in November 1952, the Terrier featured a single cylinder air-cooled ohv four stroke engine of 149 cc. The engine was built in unit with the gearbox, primary drive being by way of a single row chain housed in a separate chaincase on the left-hand side of the engine unit. The gearbox provided four ratios.

The frame featured telescopic front forks and plunger rear suspension, this combination being retained on all Terrier models. Brakes were a 5½ in diameter single leading shoe drum front and rear. Detail improvements were carried out in 1955, and again in 1956, when a larger 2¾ pint oil tank was fitted and the clutch lining material changed. Production of the T15 Terrier ceased in August 1956.

T20 Tiger Cub

The first Tiger Cub appeared in 1954, the bore and stroke of the Terrier engine having been modified to obtain a displacement of 199 cc. The gearbox ratios and Amal carburettor were modified to suit the new engine. The frame was the same as that of the Terrier, other than the adoption of 19 x 3.00 in tyres. A dualseat was fitted.

Detail changes took place in 1955, and again in 1956 when a larger petrol tank and a modified clutch were adopted. In addition, the rear of the frame was altered to accept a 3.25 x 16 in rear tyre, a similarly sized front wheel and tyre being fitted.

In October 1956, an entirely new frame was adopted, featuring swinging arm rear suspension. In September 1957, duplex primary drive was adopted, as were other minor changes including a deeper rear mudguard.

September 1958 saw the appearance of the pressed steel centre fairing along with a restyled 3 gallon petrol tank. September 1959 saw a fundamental change in the crankcases, in that they were now arranged to split vertically along the axis of the cylinder barrel mount instead of the join being well to the right as in earlier models. This change is effective from engine number 57617 onwards. Larger wheels and tyres (3.25 x 17 in) were also introduced, as was a modified cylinder head with a larger inlet tract and valve, and a larger 18 mm carburettor to suit.

In September 1960, a redesigned oil pump was incorporated, and this was again modified in September 1961, along with a new rear light unit and petrol tap. In September 1963 (engine No. 94600 onwards) a ball bearing replaced the plain timing side main bearing.

January 1966 saw the amalgamation of the BSA Bantam frame and the Tiger Cub power unit, the model being known as the Bantam Cub for obvious reasons. A new type of plunger oil pump was fitted. The standard T20 Tiger Cub was finally discontinued in December 1966.

T20C Tiger Cub Competition

Introduced in October 1956, the T20C was similar to the T20, with the exception of the following points; wide-ratio gears were employed, as were a 2.75 x 19 front tyre and a 3.50 x 18 rear tyre. A sump guard was fitted to the underside of the frame, and a high level exhaust system was used. Modifications were the same as those made to the T20 model, until September 1959, when the T20C was phased out.

T20S Tiger Cub Sports model

The T20S first appeared in October 1959, and featured modified gear ratios, heavier duty front forks of the type fitted to the 3TA and a 3.00 x 19 in front wheel and 3.50 x 18 in rear wheel. Energy transfer ignition was fitted, with direct lighting and a quickly-detachable headlamp for competition use. The model was discontinued in August 1960.

T20T Tiger Cub Trials

The successor to the T20C, the T20T was very similar to its predecessor. Production of this model ran from September 1960 to August 1961, being replaced by the TR20 Trials model.

T20 SL Tiger Cub Scrambler

The T20SL model featured a high compression engine, a larger bore Monobloc carburettor and a redesigned camshaft. Introduced in September 1960, it ran until August 1961.

T20 SS Tiger Cub Sports

This model was introduced in September 1961 as a successor to the T20 SL Scrambler. The compression ratio was raised once again, and a 2⅝ gallon petrol tank and sports mudguards were fitted.

T20 S/H Tiger Cub Sports

Introduced in 1961, this Sports Cub featured the new journal ball bearing which replaced the plain timing side main bearing. A new oil pump was fitted in September 1965, the model finally being discontinued in July of 1966.

TR20 Tiger Cub Trials

Last of the three Trials models, the TR20 was a low compression version of the T20 S/H. It employed wide-ratio gears, energy-transfer ignition, and was not equipped with lights. It ceased production in February 1965.

T20 S/C Super Cub

Introduced in November 1966, the Super Cub utilised the D7 BSA Bantam frame. It featured full-width hubs, a chromium-plated headlamp and a new tank and seat, similar to the BSA fittings. The last Cub to be produced, it ceased production in October 1968.

Ordering spare parts

Although the Triumph Engineering Company Limited is no longer in existence, many spare parts for the Terrier and Tiger Cub motorcycle are still widely available. A number of motorcycle dealers who acted as Triumph agents still retain their remaining stocks of parts. In addition, a number of specialist spare parts dealers now provide a service for owners of these machines. The carburettor and electrical components were used by a large number of manufacturers and as such are often available from specialist dealers catering for entirely different marques.

Always quote the engine and frame numbers in full, especially for these older models. Include any letters before or after the numbers itself. The importance of given this information cannot be overstressed. Quite fundamental design changes occurred when the swinging arm method of rear suspension was adopted and the engine and frame numbers provide the most positive form of correctly identifying the machine concerned. The frame number will be found stamped on the left-hand side of the gusset around the steering head, or on the lower half of the right-hand frame tube. The engine number is stamped on the left-hand crankcase, immediately below the base of the cylinder barrel.

Try to use only parts of genuine Triumph manufacture. Pattern parts are available, but in many instances they will have an adverse effect on performance and/or reliability. Some complete units were available originally on a 'Service Exchange' basis so that costs could be kept to an economic level by the supply of factory re-conditioned replacements. It is worth enquiring whether any of these facilities still exist in the form of left-over stocks on an agent's shelves or the agent's own reconditioned replacements.

Retain any broken or worn parts until a new replacement has been obtained. Often these parts are required as a pattern for identification purposes, a problem that becomes more acute when a machine is classified as obsolete. In an extreme case, where replacements are not available, it may be possible to reclaim the original or to use it as a pattern for having a replacement made. Many older machines are kept on the road in this way, long after a manufacturer's spares have ceased to be available.

Some of the more expendable parts such as sparking plugs, bulbs, tyres, oils and greases etc., can be obtained from accessory shops and motor factors, who have convenient opening hours, and can often be found not far from home. It is also possible to obtain parts on a Mail Order basis from a number of specialists who advertise regularly in the motor cycle magazines.

It is often worthwhile buying incomplete or damaged machines of the same model, since these may be purchased cheaply and will provide a good stock of otherwise unobtainable parts. The owner of any obsolete machine will very quickly begin to find the necessary contacts and dealers who will be able to assist in tracking down elusive parts.

Engine number location

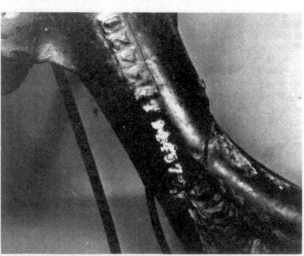

Frame number location

Routine Maintenance

Periodic routine maintenance is a continuous process that commences immediately the machine is used. It must be carried out at specified mileage recordings or on a calender date basis if the machine is not used regularly – whichever falls sooner. Maintenance should be regarded as an insurance policy rather than a chore, because it will help keep the machine in peak condition and ensure long, trouble-free service. It has the additional benefit of giving early warning of any faults that may develop and will act as a regular safety check, to the obvious benefit of both rider and machine alike.

The various maintenance tasks are described under their respective mileage and calendar date headings. Accompanying diagrams have been added, where necessary. It should be remembered that the interval between the various maintenance tasks serves only as a guide. As the machine gets older or is used under particularly arduous conditions, it would be advisable to reduce the period between each check.

Each service operation is described in detail. If additional information is required, it will be found under the relevant heading in the appropriate Chapter. No special tools are required for the normal routine maintenance tasks. The tools contained in the kit supplied with every new machine will prove adequate for each task but if they are not available, the tools found in the average household should suffice.

Weekly, or every 200 miles (320 km)

Checking engine oil level

1   Remove the oil tank filler cap and view the oil level through the orifice. Originally most oil tanks were equipped with an oil level mark transfer. If the mark is no longer visible, an estimate of the oil level height should be made. The oil level should be approximately $1\frac{1}{2}$ in (38 mm) from the filler neck. Although this level is not in itself critical, too much oil will cause the excess to be expelled via the oil tank breather. Too little oil can cause overheating, as the circulation of the oil normally provides some degree of cooling as it passes through the tank.

Tyre pressures

2   Check the pressure of both tyres, using a pressure gauge that is know to be accurate. Always check tyres when they are quite cold. After a machine has covered a number of miles, the tyres will warm up, causing the pressure to increase. The correct pressures for each model are as follows:

Front tyres:
2.75 x 19 : 16 psi (1.13 kg cm²) minimum
3.00 x 19 : 16 psi (1.13 kg cm²) minimum
3.25 x 16 : 16 psi (1.13 kg cm²) minimum
3.25 x 17 : 16 psi (1.13 kg cm²) minimum

Rear tyres:
2.75 x 19 : 18 psi (1.27 kg cm²) minimum
3.00 x 19 : 18 psi (1.27 kg cm²) minimum
3.25 x 16 : 18 psi (1.27 kg cm²) minimum
3.50 x 18 : 16 psi (1.13 kg cm²) minimum
3.25 x 17 : 18 psi (1.27 kg cm²) minimum

The above pressures may be varied to compensate for varying loads. For example, if a pillion passenger is carried, increase the front tyre pressure by 4 psi (0.28 kg cm²), and the rear tyre pressure by 6 psi (0.42 kg cm²). Always keep the valve free from road dirt, and remember to refit the valve cap to maintain a second seal in the event of valve failure. Note that machines fitted with trials or scrambles (motocross) tyres will require different pressures to suit varying applications. Different types of normal road tyres may also call for changed pressure settings. If in any doubt, the tyre manufacturer should be able to advise on the correct pressure to use.

Safety check

3   Inspect each tyre for signs of splitting, cracking or other damage. Use the blade of a screwdriver to remove any small stones or other objects which may be embedded in the treads.

Inspect the machine for loose nuts and bolts or other components, tightening where necessary.

Legal check

4   Check that all the lights are in working order and that the horn functions.

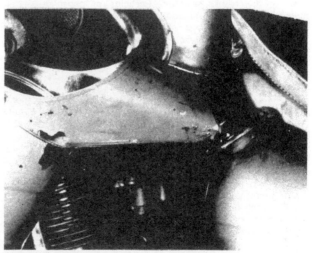

Top up oil tank to recommended level

Rear chain tension and lubrication

5   Chain adjustment is correct when there is approximately $\frac{3}{4}$ inch free play in the middle of the lower run. Always test the chain at the highest position in the rotation of the chain. Adjustment is effected by slackening the wheel spindle and torque arm nuts and moving the rear wheel backwards by means of the adjuster bolts in the fork ends. Always move the adjusters an equal amount, to maintain correct wheel alignment. When adjustment is correct, push the wheel hard against the adjuster bolts before tightening the spindle. Do not omit to tighten the torque arm retaining nut.

If there is any doubt about the accuracy of wheel alignment, this can be checked by using a long straight edged plank positioned along the machine.

Lubricate the chain using engine oil or one of the proprietary aerosol lubricants. The latter type is more efficient as it is less prone to being flung off by the fast rotating chain.

Monthly, or every 1,000 miles (1600 km)

Complete the maintenance tasks listed under the preceding weekly heading, then the following additional items:

Battery electrolyte level

1   Remove the dualseat and detach the battery cover by removing the strap. Check that the electrolyte level in each cell is approximately $\frac{1}{4}$ inch above the plates. Replenish, if necessary, with distilled water. Do not use tap water which contains many mineral impurities which will reduce the efficiency of the battery.

Changing the engine oil

2   The interval specified for engine oil changes by Triumph is 1,500 miles, but it is worthwhile reducing this to every 1000 miles in view of the fact that even the latest models will by now be at least ten years old, and engine spares are likely to become progressively harder to obtain. Where the engine has just been rebuilt, intermediate 250 and 500 mile oil changes should also be carried out as the initial 1000 miles of running will produce much more foreign matter than is usual.

Oil changes should always be carried out whilst the engine is warm. This ensures that the abrasive particles in the oil system are in suspension in the oil, and that the warm oil will drain more quickly and completely than when cold.

Remove the crankcase oil filter by unscrewing the large hexagon-headed sump plug. A certain amount of residual oil will be released and should be caught in a drain tray. The oil tank contents should be drained after removing the drain plug, noting that the tank will contain approximately $2\frac{1}{2}$ pints of oil. Remove the oil tank filler, and wash this and the crankcase filter in clean petrol to remove accumulated sediment. During cleaning, note the composition of this sediment, particularly in the case of the crankcase filter. Particles of phosphor-bronze, aluminium alloy or steel, especially if numerous and unusually large, indicate that a bush, bearing or some other component has started to disintegrate. This condition warrants immediate investigation and rectification, if severe engine damage is to be avoided. On no account leave it in the hope that it will go away – it will only get worse.

Periodically, it is well worth removing the oil tank from the machine and flushing it thoroughly with solvent flushing oil or a mixture of paraffin (kerosene) and oil. The oil feed pipe should be cleaned and blown out with compressed air during this operation.

When reassembling the various plugs and filters, ensure that the sealing washers are in good condition and renew as necessary. Most motorcycle dealers sell small packs of fibre sealing washers, though copper washers, where used, are rather harder to obtain.

Refill the oil tank to within $1\frac{1}{2}$ inch of the filler neck using the correct monograde oil. In the Summer months, SAE 30 should be used, whilst SAE 20 is recommended for Winter use.

It is not recommended that multigrade oils be used in this engine.

IMPORTANT NOTE: Certain competition engines may have been run on a vegetable-based oil such as Castrol R. These oils are **not** compatible with mineral oils, and if mixed will rapidly form a rubbery sludge in the oilways, causing lubrication failure. If it is wished to change from one type to the other, the engine must be stripped and all traces of the old oil removed from the engine components.

Changing the primary chaincase oil

3   Remove the drain plug in the lower edge of the chaincase, and allow the oil to drain into a suitable container. The chaincase holds about $\frac{1}{3}$ pint (200 cc) of oil. Refit and tighten the drain plug and remove the level plug and filler plug from the front of the case. Add SAE 20 engine oil until it just begins to seep out of the level hole, then replace the two plugs. Check that the chaincase securing screws are tight each time the oil is changed.

Checking the gearbox oil level

4   The drainplug on the underside of the gearbox is hollow, and is fitted with a smaller hexagon headed plug. This latter plug screws into the end of a short tube which projects up into the gearbox and corresponds with the correct oil level. Remove the smaller level plugs and the gearbox filler plug (the rearmost of the two slotted plugs on the top of the crankcases). Add SAE 30 engine oil until it begins to seep out of the level hole in the drain plug. Refit and tighten the level plug.

Rear chain lubrication

5   Lubrication of the rear chain may be carried out with the chain in place on the machine, as described under the weekly/200 mile heading. At greater intervals, however, the chain should be removed from the machine, cleaned thoroughly and relubricated by immersion in a hot bath of special chain grease such as 'Linklyfe or 'Chainguard'. To remove the chain, disconnect it at the spring link. If possible, connect an old chain to one end of the chain, which can then be pulled into place on the gearbox sprocket. By using this method the relubricated chain can be pulled back into position with ease.

General lubrication

6   Apply a grease gun to the rear suspension plungers, or the swinging arm pivot grease nipples. Grease also the wheel hubs and brake cam spindles, if grease nipples are provided.

Note that both these points should be greased sparingly, to prevent grease from entering the brake drums.

Use aerosol lubricant on rear chain at frequent intervals

Top up gearbox until oil drips from level plug

Check valve clearances after removing inspection covers

Three monthly or every 3000 miles (5000 km)

Complete the tasks listed under the preceding headings, then carry out the following:

Valve clearance adjustment

1   Valve clearance adjustment should be carried out every 3000 miles or at any time the valve operation becomes noisy. Start by placing the machine on the centre stand so that the rear wheel is raised clear of the ground and is free to rotate. Select top gear, remove the sparking plug and detach the rocker box covers. It should be noted that the clearance should be adjusted only when the engine is cold, preferably after standing overnight.

Turn the engine over by way of the rear wheel until the inlet valve is just closing. Insert a piece of wire into the sparking plug hole so that the top of the piston can be felt. Continue turning the wheel slightly until the piston reaches the top of its stroke (TDC). At this position (Top Dead Centre on the compression stroke) both valves will be fully closed and the clearance in each valve train at maximum.

It is easier to set clearances than to check them, so slacken off each locknut and adjuster screw, insert a feeler gauge of the correct thickness, then tighten the screw so that it just nips the feeler guage. Holding the adjuster in this position, tighten the locknut and then recheck the clearance. Repeat on the remaining rocker.

When refitting the rocker covers, ensure that the gaskets and copper or fibre washers are in good condition.

The correct valve clearances are as follows:

T.15 Terrier and T.20 Standard Tiger Cub models
  Inlet   0.010 in (0.254 mm) *(camshaft No. E4869)*
  Exhaust 0.010 in (0.254 mm)
T.20SS, T.20SH, T.20SL, some T20S models and any machine fitted with the sports camshaft.
  Inlet   0.002 in (0.05 mm) *(camshaft No. E4870 or*
  Exhaust 0.004 in (0.10 mm)   E3183*

It will be seen from the above figures that it is important that the tappet clearance is set to match the appropriate camshaft. Normally, this can be checked with the previous owner when buying the machine, but if for any reason the proper setting is not known, set the clearances as per the standard camshaft at 0.010 in. If when started the engine proves excessively noisy (this should be immediately apparent) it would probably indicate that the sports cam is fitted. On no

account must the standard camshaft models be run with the smaller valve clearances.

Contact breaker gap

2   All Terrier models, and early Tiger Cub models were equipped with a distributor mounted vertically on the right-hand crankcase half, just behind the cylinder barrel. On later models, the distributor was omitted, and its mounting hole plugged. The contact breaker assembly, along with the automatic timing unit (ATU) was relocated in the front of the right-hand casing.

Remove the contact breaker assembly cover and turn the engine until the points are in the fully open position. Clean the points faces using a fine swiss file or a strip of emery paper (No. 400) backed by a thin strip of tin. If the points faces are badly blackened, burned or pitted, they should be removed for further attention as described in Chapter 3.

Using a feeler gauge, check that the gap between the points is within the range 0.014 – 0.016 in (0.35 – 0.40 mm). This check must be made when the points are in the fully open position. Ideally, a 0.015 in feeler gauge should be a light sliding fit between the contact faces. Adjustment may be carried out by slackening the fixed contact securing screw just enough to permit its movement. Set the gap, then retighten the securing screw. Remember to re-check the setting before replacing the cover, to ensure that the fixed contact was not moved when the screw was retightened. A drop of light machine oil should be applied to the felt lubricating wick to prevent premature wear of the contact breaker fibre heel.

Ignition timing

3   It is advisable to check that the ignition system is accurately timed when the contact breaker points receive attention. Details of this operation are given in Chapter 3. Pay particular attention to the General Notes on ignition timing – these will save much unnecessary work during this operation.

Sparking plug

4   Remove the sparking plug and clean the area around the electrodes with a wire brush, then attend to the points gap with fine emery paper. Set the gap to 0.020 in (0.50 mm) by bending the outer electrode nearer to or further away from the centre electrode. Before replacing the plug, lubricate the threads with graphite grease, to aid future removal.

Six monthly, or every 5,000 miles (8000 km)

Again complete all the routine maintenance tasks listed previously, then the following additional tasks:

Gearbox oil change

1   Place a suitable container below the gearbox and remove the drain plug. The gearbox oil must be changed when the oil is warm, to improve the flow. Replenish the gearbox, with SAE 30 engine oil, after replacing the drain plug. The gearbox capacity is $\frac{1}{3}$ pint. Do not overfill the gearbox or the oil will find its way along the mainshaft into the clutch.

Air filter element

2   Two basic types of air filter element were employed on Tiger Cub models, both serving to prevent the ingress of abrasive dust into the engine. The first type is of the oil-impregnated gauze type, and is normally mounted on the end of the carburettor. This type can be removed and washed thoroughly in clean petrol to remove all traces of road dirt, and then blown dry with compressed air or left, preferably overnight, to let the petrol evaporate. It is a good idea to wet the gauze with engine oil before the filter is refitted. This will help to trap any dust particles drawn through it.

A second type of filter, using a corrugated paper element, was used on some machines. In this instance, the element is incorporated in or positioned near to the battery and toolbox casing. The paper element should not be washed out, as this will only succeed in obstructing the porous surface. Loose dirt can be dislodged by tapping the element on a hard surface, or better still, by blowing compressed air through it from the inside face. If the filter becomes damp, contaminated with oil, or perforated, it should be renewed. It should be noted that it may be possible to adapt a modern car-type element if the original type proves impossible to obtain.

Note that if the engine is run without the correct type of filter in position, the mixture strength will be upset, and may result in the engine overheating. It is also likely that the rate of engine wear will be significantly increased due to dust being drawn into the cylinder.

Yearly, or every 10,000 miles (16000 km)

After completing the weekly, three monthly and six monthly tasks, continue with the following additional items:

All these items require greater attention than the previously described shorter interval maintenance operations. For full details refer to the relevant Chapters within the main text.

Drain and replenish the fork oil. Check the fork bushes and oil seals for wear, renewing where considered necessary.

Remove the wheels and check the condition of the brake linings. If necessary, renew the lining and/or shoes.

Remove the wheel bearings and clean thoroughly. Check the bearings and renew if worn. Repack with grease on replacing.

Detach and dismantle the carburettor. Clean all the components and airways, replacing any jets if worn.

Repack the steering head bearings with grease after removal for cleaning.

Remove the cylinder head and give a complete top end overhaul.

Other adjustments and checks

Brake adjustment

1   The interval at which the brakes will require adjustment depends largely on the purpose to which the machine is put and the manner in which it is ridden. It follows that a machine consistently ridden hard will suffer more rapid brake lining wear than one used for commuting only.

Adjustment of the front brake is made by screwing the cable adjuster in or out until a small amount of play is present before the brake comes into operation. The rear brake is of rod operation. It is adjusted by a knurled nut on the rod. The amount of travel of the brake pedal should be approximately two inches before the brake is applied fully.

Clutch adjustment

2   In common with the brakes, clutch wear and the necessity for adjustment depends upon the manner in which the machine is used. Clutch adjustment should always be carried out in two stages, as follows:

Disconnect the clutch cable at the handlebar lever to ensure that the maximum amount of cable free play is available. With the primary chaincase inspection screw removed, the adjuster and locknut in the centre of the clutch pressure plate will be visible. Slacken the locknut, and screw the adjuster inwards until the clutch actuating lever touches the inside of the right-hand outer casing, at which point a positive stop will be felt. Slacken the adjuster by $\frac{1}{2}$ turn, and hold it whilst the locknut is tightened. The inspection cap may now be replaced. The cable should be re-connected at the handlebar lever, and adjusted to give $\frac{1}{16}$ in (1.5 mm) free movement.

Control cable lubrication

3   Cable life, and efficiency, are dependent on regular lubrication. The intervals at which this should take place are governed by the general climatic conditions, but as a general rule, any obvious grime or stiffness in operation, warrants cleaning and lubrication.

Use motor oil or an all-purpose oil to lubricate the control cables. A good method for lubricating the cables is shown in the accompanying illustration, using a plasticine funnel. This method has a disadvantage in that the cables usually need removing from the machine. An hydraulic cable oiler which pressurises the lubricant, overcomes the problem. Do not lubricate nylon lined cables as the oil will cause the nylon to swell, thereby causing total cable seizure.

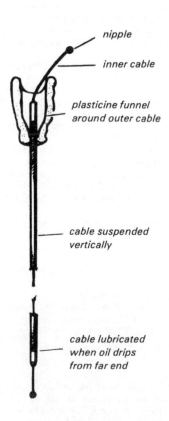

nipple

inner cable

plasticine funnel around outer cable

cable suspended vertically

cable lubricated when oil drips from far end

Control cable oiling

Gauze elements should be removed and washed in petrol

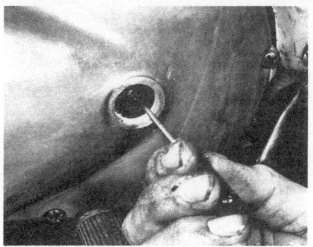

Clutch pushrod clearance is set via inspection hole

Recommended lubricants and adjustment settings

Engine (oil tank)
    Summer: SAE 30 engine oil   About $2\frac{3}{4}$ imp pints (1.6 litres, 1.7 US
    Winter: SAE 20 engine oil   pints) or to $1\frac{1}{2}$ in of filler neck

Gearbox   SAE 30 engine oil $\frac{1}{3}$ imp pint (200 cc, 0.8 US pints)

Primary chaincase   SAE 20 engine oil $\frac{1}{3}$ imp pint (200 cc, 0.8 US pints)

Front forks   Fork oil or SAE 30 engine oil $\frac{1}{8}$ imp pint (75 cc, 0.3 US pints) per leg

Wheel bearings   High melting point grease   As required

Lubrication points
    Grease nipples
    Brake fulcrums   High melting point grease   As required
    Stand and pedal pivots

Final drive chain
    Intermediate lubrication   Aerosol chain lubricant
    Full lubrication   Hot immersion type chain lubricant

Contact breaker wick   Light machine oil   1 – 2 drops

Contact breaker gap   0.014 – 0.016 in (0.36 – 0.40 mm)

Sparking plug gap   0.020 in (0.50 mm)

Sparking plug type
    T20S model   Champion L-5 or Motorcraft AE3
    All other models   Champion L-7 or L-85 or Motorcraft AE3

Working conditions and tools

When a major overhaul is contemplated, it is important that a clean, well-lit working space is available, equipped with a workbench and vice, and with space for laying out or storing the dismantled assemblies in an orderly manner where they are unlikely to be disturbed. The use of a good workshop will give the satisfaction of work done in comfort and without haste, where there is little chance of the machine being dismantled and reassembled in anything other than clean surroundings. Unfortunately, these ideal working conditions are not always practicable and under these latter circumstances when improvisation is called for, extra care and time will be needed.

The other essential requirement is a comprehensive set of good quality tools. Quality is of prime importance since cheap tools will prove expensive in the long run if they slip or break and damage the components to which they are applied. A good quality tool will last a long time, and more than justify the cost. The basis of any tool kit is a set of open-ended spanners, which can be used on almost any part of the machine to which there is reasonable access. A set of ring spanners makes a useful addition, since they can be used on nuts that are very tight or where access is restricted. Where the cost has to be kept within reasonable bounds, a compromise can be effected with a set of combination spanners – open-ended at one end and having a ring of the same size on the other end. Socket spanners may also be considered a good investment, a basic $\frac{3}{8}$ in or $\frac{1}{2}$ in drive kit comprising a ratchet handle and a small number of socket heads, if money is limited. Additional sockets can be purchased, as and when they are required. Provided they are slim in profile, sockets will reach nuts or bolts that are deeply recessed. When purchasing spanners of any kind, make sure the correct size standard is purchased. Almost all machines manufactured outside the UK and the USA have metric nuts and bolts, whilst those produced in Britain have BSF or BSW sizes. The standard used in the USA is AF, which is also found on some of the later British machines. Other tools that should be included in the kit are a range of crosshead screwdrivers, a pair of pliers and a hammer.

When considering the purchase of tools, it should be remembered that by carrying out the work oneself, a large proportion of the normal repair cost, made up by labour charges, will be saved. The economy made on even a minor overhaul will go a long way towards the improvement of a tool kit.

In addition to the basic tool kit, certain additional tools can prove invaluable when they are close to hand, to help speed up a multitude of repetitive jobs. For example, an impact screwdriver will ease the removal of screws that have been tightened by a similar tool, during assembly, without a risk of damaging the screw heads. And, of course, it can be used again to retighten the screws, to ensure an oil or airtight seal results. Circlip pliers have their uses too, since gear pinions, shafts and similar components are frequently retained by circlips that are not too easily displaced by a screwdriver. There are two types of circlip pliers, one for internal and one for external circlips. They may also have straight or right-angled jaws.

One of the most useful of all tools is the torque wrench, a form of spanner that can be adjusted to slip when a measured amount of force is applied to any bolt or nut. Torque wrench settings are given in almost every modern workshop or service manual, where the extent is given to which a complex component, such as a cylinder head, can be tightened without fear of distortion or leakage. The tightening of bearing caps is yet another example. Overtightening will stretch or even break bolts, necessitating extra work to extract the broken portions.

As may be expected, the more sophisticated the machine, the greater is the number of tools likely to be required if it is to be kept in first class condition by the home mechanic. Unfortunately there are certain jobs which cannot be accomplished successfully without the correct equipment and although there is invariably a specialist who will undertake the work for a fee, the home mechanic will have to dig more deeply in his pocket for the purchase of similar equipment if he does not wish to employ the services of others. Here a word of caution is necessary, since some of these jobs are best left to the expert. Although an electrical multimeter of the AVO type will prove helpful in tracing electrical faults, in inexperienced hands it may irrevocably damage some of the electrical components if a test current is passed through them in the wrong direction. This can apply to the synchronisation of twin or multiple carburettors too, where a certain amount of expertise is needed when setting them up with vacuum gauges. These are, however, exceptions. Some instruments, such as a strobe lamp, are virtually essential when checking the timing of a machine powered by a CDI ignition system. In short, do not purchase any of these special items unless you have the experience to use them correctly.

Although this manual shows how components can be removed and replaced without the use of special service tools (unless absolutely essential), it is worthwhile giving consideration to the purchase of the more commonly used tools if the machine is regarded as a long term purchase. Whilst the alternative methods suggested will remove and replace parts without risk of damage, the use of the special tools recommended and sold by the manufacturer will invariably save time.

Chapter 1 Engine, clutch and gearbox

Contents

General description 1
Operations with the engine in the frame 2
Operations with the engine removed 3
Method of engine removal 4
Dismantling the engine: general 5
Removing the cylinder head, barrel and piston 6
Dismantling the alternator rotor and primary drive 7
Removing the distributor 8
Removing the contact breaker plate and automatic
timing unit (September 1963 onwards) 9
Removing the inner timing cover and gearbox
components 10
Removing the camshaft, tappets and oil pump 11
Removing the crankshaft pinion assembly 12
Crankcase types: identification and construction 13
Separating the crankcase halves: general note 14
Separating one-piece crankcase halves 15
Separating two-piece crankcase halves 16
Examination and renovation: general 17
Big end and main bearings: examination and renovation .. 18
Small end bush: examination and renovation 19
Timing pinion and camshaft: examination and renovation . 20
Cylinder barrel: examination and renovation 21
Piston and piston rings: examination and renovation 22

Valves, valve springs and valve guides: examination and
renovation 23
Cylinder head: examination and renovation 24
Rockers, rocker spindles and rocker covers: examination
and renovation 25
Gearbox components and kickstart assembly: examination
and renovation 26
Clutch: examination and renovation 27
Engine reassembly: general 28
Engine reassembly: fitting the mainshaft sleeve gear and
gearbox sprocket 29
Engine reassembly: joining the crankcase halves 30
Engine reassembly: fitting the timing pinion, tappets,
camshaft and oil pump 31
Engine reassembly: fitting the gearbox components, inner
casing, kickstart assembly and outer cover 32
Engine reassembly: refitting the clutch and primary drive . 33
Engine reassembly: refitting the piston and cylinder
barrel 34
Engine reassembly: setting the ignition timing 35
Engine reassembly: refitting the cylinder head 36
Fitting the engine unit into the frame 37
Final adjustments 38
Starting and running the rebuilt engine 39

Specifications

Engine

Type .... Single cylinder, air cooled, four-stroke
Cylinder barrel .... cast iron

	Terrier	Tiger Cub
Bore	57.0 mm (2.24 in)	63.0 mm (2.48 in)
Stroke	58.5 mm (2.30 in)	64.0 mm (2.52 in)
Displacement	149 cc (9.09 cu in)	199 cc (12.14 cu in)

bhp:
   T.15 Terrier .... 8.3 @ 6500 rpm, 8.0 @ 6000 rpm
   T.20, T.20C and T.20T .... 10 @ 6000 rpm
   T.20SS, T.20SH and T.20S .... 14.5 @ 6500 rpm
Torque (T.15 only) .... 185 lb in @ 5000 rpm, 150 lb in @ 2750 rpm
Compression ratio:
   T.15, T.20, T.20C, T.20T .... 7 : 1
   T.20SS, T.20SH, T.20S .... 9 : 1

Piston
    Type . . . .   Aluminium alloy

Piston rings
    Number . . . .   Two compression, one oil control

End gap:	Terrier	Tiger Cub
Maximum	0.008 in (0.20 mm)	0.010 in (0.24 mm)
Minimum	0.006 in (0.15 mm)	0.008 in (0.20 mm)

Valve clearances
  T.15, T.20 and any model using standard camshaft (No. E4869)
Inlet	0.010 in (0.25 mm)
Exhaust	0.010 in (0.25 mm)

  T.20SS, T.20SH, T.20SL, some T.20S models and any machine fitted with sports camshaft (No. E4870 or E3183)
Inlet	0.002 in (0.05 mm)
Exhaust	0.004 in (0.10 mm)

Valve timing
  T.15 Terrier, T.20, T.20C, T.20T, all standard camshaft models:
Inlet opens at	30° BTDC
Inlet closes at	50° ABDC
Exhaust opens at	55° BBDC
Exhaust closes at	25° ATDC

  T.20S, T.20SS, T.20SH, all sports camshaft models:
Inlet opens at	39° BTDC
Inlet closes at	61° ABDC
Exhaust opens at	65° BBDC
Exhaust closes at	35° ATDC

Sprocket sizes

	Engine	Clutch	Gearbox	Rear wheel
T.15	19T	48T	17T	48T
T.20 up to 17388*	19T	48T	18T	48T
T.20 from 17388 to 35846*	18T	36T	17T	54T
T.20 from 35846 to 56360*	19T	48T	18T	46T
T.20C up to 35846*	18T	36T	16T	54T
T.20C from 35847*	19T	48T	16T	46T
T.20 from 56360*	19T	48T	17T	46T
T.20T	19T	48T	16T	54T
T.20S	19T	48T	17T	48T
T20SS	–	–	17T	48T
T20SH	–	–	17T	48T

Chains

	Primary	Secondary
T.15, T.20 up to 17388*	3/8 x 7/32 in 62 rollers (single row)	1/2 x 3/16 in 112 rollers
T.20 from 17388 to 35846*	1/2 x 3/16 in 48 rollers (single row)	1/2 x 3/16 in 116 rollers
T.20 from 35846 to 56360*	3/8 in 62 rollers (duplex)	1/2 x 3/16 in 112 rollers
T.20C up to 35846*	1/2 x 3/16 in 48 rollers (single row)	1/2 x 3/16 in 116 rollers
T.20C from 35847*, T.20 from 56360*	3/8 in duplex 62 rollers	1/2 x 3/16 in 112 rollers
T.20T	3/8 in duplex 62 rollers	1/2 x 3/16 in 116 rollers
T.20S	3/8 in duplex 62 rollers	1/4 x 3/16 in 113 rollers
Late T.20	3/8 in duplex 62 rollers	5/8 x 3/8 in 112 rollers
T.20SS, T.20SH	3/8 in duplex 62 rollers	5/8 x 3/8 in 113 rollers

* engine number

Gearbox ratios

	1st	2nd	3rd	Top
T.15 Terrier	21.1 : 1	14.8 : 1	9.5 : 1	7.1 : 1
T.20 up to 17388*	20.0 : 1	13.8 : 1	8.8 : 1	6.7 : 1
T.20 from 17388 to 35846*	19.0 : 1	13.1 : 1	8.3 : 1	6.35 : 1
T.20 from 35846 to 56360*	19.4 : 1	13.3 : 1	8.5 : 1	6.45 : 1
T.20C up to 35846*	20.0 : 1	13.8 : 1	8.8 : 1	6.7 : 1
T.20C from 35847*	21.5 : 1	14.8 : 1	9.5 : 1	7.26 : 1
T.20 from 56360*	20.3 : 1	14.0 : 1	9.0 : 1	6.84 : 1
T.20T	28.1 : 1	19.4 : 1	12.4 : 1	8.55 : 1
T.20S, T.20SH, T.20SS	19.8 : 1	13.4 : 1	8.6 : 1	7.13 : 1

* engine number

Clutch

No. of plates	
Plain	3, plus clutch cover
Friction	3, plus clutch drum/chain wheel
Nominal friction plate thickness	$\frac{1}{8}$ in (3.18 mm)
Lining thickness	$\frac{1}{32}$ in (0.79 mm)
Clutch springs	3
Free length	
$1\frac{21}{32}$ (42 mm)	nominal
$1\frac{17}{32}$ (39 mm)	wear limit

1 General description

The T15 Terrier and the T20 Tiger Cub models are equipped with a single cylinder, air-cooled four stroke engine of 149 cc and 199 cc respectively. The engine and gearbox are of unit construction, the gearbox components being housed in a separate compartment to the rear of the casing halves. The crankcases are jointed vertically, in a fore and aft plane, and support the crankshaft which comprises a crankpin and connecting rod, two full flywheels and two mainshafts. The early models, up to September 1959, featured the main part of the crankcase being incorporated in the right-hand casing half, the primary chaincase forming a 'lid' over the open end of the chamber. Models after this date utilised different cases which separated along the centre axis.

The crankshaft assembly is supported by two main bearings; on all models the drive side (left-hand) main bearing is of the journal ball type. On machines up to engine number 94600 (September 1963) a plain timing side bearing, or bush, was utilised. After this point, a second journal ball bearing was used.

A light alloy piston is employed, running in a cast-iron cylinder barrel, the latter being retained by four holding down studs, which also retain the aluminium alloy cylinder head. Overhead valves are fitted to the cylinder head. These are operated by rocker arms, which in turn are actuated by two pushrods. The pushrods are housed in a chromium plated tube on the right-hand side of the cylinder barrel. They convey movement to the valve rockers, from the tappets and camshaft.

All models are fitted with a coil ignition system, powered by a 6 volt battery, which in turn is fed by the crankshaft-mounted alternator. The timing of the ignition spark is controlled by a contact breaker set, mounted in a distributor unit on pre-1963 models. On later models, the distributor arrangement was abandoned, and a contact breaker assembly housed in the right-hand outer cover.

Carburation was by way of a Zenith instrument on machines manufactured before September 1961. Subsequent models were equipped with an Amal Type 32 carburettor, whilst some later competition machines used an Amal Monobloc unit.

2 Operations with the engine in the frame

1 It is not necessary to remove the engine from the frame unless the crankshaft assembly requires attention. Most operations can be accomplished with the engine in the frame, such as:

   a) *Removal and replacement of the cylinder head.*
   b) *Removal and replacement of the cylinder barrel and piston.*
   c) *Removal and replacement of the clutch and primary drive.*
   d) *Removal of the timing pinions.*
   e) *Removal of the oil pump.*
   f) *Removal of the alternator.*
   g) *Removal of the gearbox components.*

2 When several operations have to be undertaken, such as during an extensive rebuild or overhaul, it is often advantageous to remove the engine from the frame after some preliminary dismantling. This will give the advantage of better access and more working space, especially if the engine is attached to a bench-mounted stand.

3 Operations with the engine removed

1 Removal and replacement of the main bearings.
2 Removal and replacement of the crankshaft assembly

4 Method of engine removal

1 Place the machine on its centre stand, choosing a position which allows plenty of working room on both sides. The work can be made much easier by placing the machine on some kind of raised platform. An old, stout, table with the legs cut down is ideal for this purpose.
2 Start by turning off the petrol tap and disconnecting the feed pipe, either at the tap, or by releasing the union at the carburettor. Slacken the top rear suspension mounting bolts, and the single front bolt, to allow the seat to be disengaged. Remove the petrol tank mounting bolts, and lift the tank clear, placing it somewhere safe to await reassembly.
3 If the unit is to be completely dismantled, the chaincase, gearbox and oil tank contents should be drained into an old plastic bowl or similar. The various drain plugs should be refitted once the oil has finished draining out. On machines fitted with a rear fairing or enclosure, this should be removed after releasing the retaining screws. Where a battery is fitted, this should be disconnected after removing the left-hand toolbox lid. It is not necessary to remove the complete chainguard, but where a short front chainguard section is employed, this should be released.
4 Turn the rear wheel to position the joining link on the rear wheel sprocket, then prise off the retaining clip and slide the chain apart. Reassemble the joining link on one end of the chain, then draw it clear of the gearbox sprocket. Disconnect the rear brake operating rod from the pedal, after removing the split pin which retains it. Slacken and remove the brake pedal retaining nut, and pull the pedal off its pivot. The front footrest mounting nuts should also be removed, and the footrests detached.
5 On models equipped with a paper element air filter, disconnect the rubber hose from the carburettor. Release the carburettor top and withdraw the valve assembly and spring. On Amal instruments, the top is retained by a knurled, threaded ring, whilst on Zenith units, it is held by a single screw. The assembly should be positioned clear of the engine unit by lodging it across the top frame tube, leaving the carburettor body undisturbed.
6 Disconnect the oil tank feed and scavenge pipes at the crankcase union block. This is retained by a single stud, which when the nut and washer are removed, will allow the pipes to be swung clear of the engine. Place a piece of rag beneath the union block to catch any residual oil which may drain out of the oil pipes. Note that it may prove necessary to remove the lower engine bolt to obtain sufficient clearance for the pipes to be disconnected.

7   Slacken the rocker feed pipe union nuts, holding the union with pliers or a self-grip wrench to prevent the copper pipe from fracturing. Remove the rocker feed carefully, taking care not to lose the copper sealing washers, as these may be re-used. Lodge the feed pipe assembly along the top frame tube, where it will not foul the unit during removal.

8   Slacken the clamp bolt which passes through the finned exhaust pipe clamp, and slide the latter away from the cylinder head. It may be helpful to spread the clamp slightly by inserting a screwdriver blade in the gap. Release the silencer mounting bolt, which on most models incorporates the rear footrest assembly on the right-hand side. The complete system may now be detached and placed to one side. Note that on certain high-level exhaust pipes there is a supplementary bracket, at the front of the exhaust pipe, which must also be removed.

9   Trace the alternator output leads, and disconnect them at the bullet connectors. On machines equipped with an ignition coil mounted in a position likely to impede engine removal, it should be detached to provide clearance; otherwise, the high and low tension terminals should be released.

10 Disconnect the clutch cable at the operating lever, then release the lower end at the gearbox. On early models, it will be necessary to remove the kickstart and gearchange pedals, and the right-hand outer cover before the cable can be disengaged from the actuating lever. On later models, a plastic plug can be prised out of the cover to permit the cable to be released without the need to remove it.

11 The engine and gearbox unit can be removed by one person relatively easily, although it is useful to have a second pair of hands to help steady the frame whilst the unit is manoeuvred out of position. Slacken and remove the engine mounting bolts and nuts, leaving the unit resting on the cradle tubes. Once the bolts are all clear, lift the front of the engine so that the crankcase lug clears the front bracket, at the same time dropping the back of the unit so that the rear lug corresponds with the cutaway provided. The unit will now move rearwards by about $\frac{1}{8}$ in, when it should be possible to clear the front lug completely. The complete engine/gearbox unit can now be lifted out from the left-hand side and placed on a bench to await further dismantling.

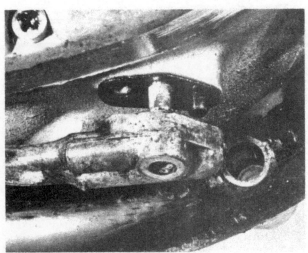

4.6 Release the oil pipe manifold from underside of crankcase

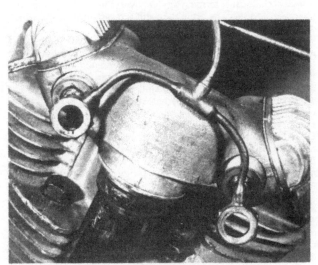

4.7 Rocker feed pipes are retained by two domed nuts

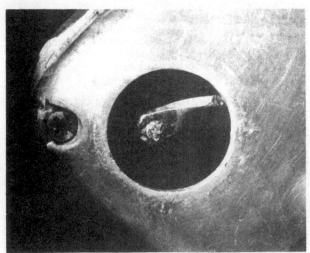

4.10 Unhook clutch cable at actuating arm

4.11a Engine unit is easily removed by one person ...

4.11b ... and can be placed on workbench for dismantling

5 Dismantling the engine: general

1   Before commencing work on the engine unit, the external surfaces should be cleaned thoroughly. A motorcycle engine has very little protection from road grit and other foreign matter, which will find its way into the dismantled engine if this simple precaution is not observed. One of the proprietary cleaning compounds, such as Gunk, can be used to good effect, particularly if the compound is allowed to work into the film of oil and grease before it is washed away. When washing down, make sure that water cannot enter the carburettor or the electrical system, particularly if these parts have been exposed.

2   Never use undue force to remove any stubborn part, unless mention is made of this requirement. There is invariably good reason why a part is difficult to remove, often because the dismantling operation has been tackled in the wrong sequence. Dismantling will be made easier if a simple engine stand is constructed that will correspond with the engine mounting points. This arrangement will permit the complete unit to be clamped rigidly to the workbench, leaving both hands free.

6 Removing the cylinder head, barrel and piston

1   The cylinder head, barrel and piston may be removed with the unit in the frame, provided the following point is noted; on machines prior to engine No: 57617 (pre-September 1959), it is necessary to remove the front and lower engine mounting bolts so that the unit can tip forwards to provide sufficient clearance. On later models, the holding down studs were shortened and the nuts lengthened by a similar amount, in which case the work may be undertaken without resorting to the above measures. The exhaust pipe, carburettor and rocker oil feed must first be detached in both cases.

2   Slacken the rocker cover retaining nuts and remove the sparking plug, as these parts may prove difficult to detach at a later stage. Slacken the four cylinder head nuts using a $\frac{1}{4}$ in Whitworth ring spanner. It is not normally difficult to dislodge the cylinder head, using a hide mallet if it proves stubborn. Resist the temptation to prise the joint apart with a screwdriver, as this invariably results in marred or damaged cylinder head or barrel fins. The pushrod tube (or tunnel), and the two pushrods, will be freed as the cylinder head is lifted away. These can be removed and placed to one side. It is worth marking the pushrods to denote inlet and exhaust, by using a piece of insulating tape or masking tape, as it is preferable to refit these in the same position as they were before removal. Note that the inner item is the exhaust pushrod.

3   The cylinder barrel may now be slid upwards off the holding down studs. Position the piston at TDC, and pack the crankcase mouth with clean rag before the piston emerges from the bore. This will prevent any debris from falling into the crankcase in the event of there being broken rings.

4   With the cylinder barrel removed, the piston may be released from the connecting rod. Prise out the old circlips and discard them. On no account should the circlips be retained and reused, as there is a real danger of their becoming displaced and causing extensive engine damage. The gudgeon pin can now be displaced, allowing the piston to be lifted away. If tight, the pin may be tapped out of position, provided that the piston is supported to avoid straining the connecting rod or bearings. A better method is to warm the piston with a rag soaked in hot water. This will expand the light alloy bosses, allowing the steel pin to push out easily.

5   Mark the inside of the piston to denote front, so that it is replaced in the same position. Obviously, if the engine is to be rebored, this step can be ignored.

5.1 Right-hand view of complete engine/gearbox unit

6.2a Lift the cylinder head clear of holding down studs

6.2b Pushrods and tube can be lifted away as a unit

6.3 Cylinder barrel can now be lifted away

6.5 Prise out circlips and displace gudgeon pin to release piston

7 Dismantling the alternator rotor and primary drive

1   The primary drive and alternator can be dismantled with the engine unit in the frame, once the oil content of the chaincase has been drained, and the brake pedal and left-hand footrest have been detached.

2   Slacken and remove the nine chaincase screws, then lift the cover away from inner casing half. As the cover comes away, it will be necessary to guide the alternator output leads through the hole in the inner casing face. Note that these must be disconnected at the bullet connectors if the unit is still in the frame. It will be noticed that the stator assembly remains in place on the inside of the cover.

3   It will be necessary to prevent crankshaft rotation whilst the alternator securing nut is slackened. If the engine unit is in the frame, this may be achieved by selecting top gear and applying the rear brake. If, on the other hand, the unit has been removed and is being dismantled, a small bar may be passed through the small end eye, and arranged to rest on wooden blocks placed each side of the crankcase mouth.

4   With the crankshaft effectively locked, the alternator rotor securing nut may be removed. Knock back the tab washer, and slacken the nut with a ring spanner or socket. Note that it may

prove necessary to strike the spanner to jar the nut free. The alternator rotor may now be pulled off the crankshaft, noting that on some models two keyways are fitted. If this is the case, mark the relevant keyway to aid reassembly. The rotor is followed by a spacer, and on some models, a washer. The former may be dislodged by levering it off with a pair of screwdrivers. On some models, the spacer has a shouldered end. Place the rotor and its related parts inside the stator for safe keeping, taking care that the Woodruff key is not misplaced.

5   Remove the clutch pressure screws using the key provided in the toolkit or a modified screwdriver. Note that some resistance will be met as the screw heads pass the end of the spring coils, due to a moulded pip on the underside of the head which prevents the screws from slackening during normal use. A small screwdriver can be used to depress the spring while this point is passed.

6   With the three screws removed, the clutch pressure plate may be lifted away, complete with the spring cups and springs. Pull out the clutch plates, using a small screwdriver or a length of stiff wire to dislodge them. The clutch housing/chainwheel can now be drawn off, together with the primary chain and engine sprocket. Note that the housing bearing contains sixteen $\frac{5}{32}$ in balls, which may drop free if the bearing ring is displaced.

7   The clutch centre is keyed to a taper and is retained by a nut and lockwasher. It is necessary to prevent its rotation whilst the nut is slackened, either by selecting top gear and applying the rear brake, if the unit is in the frame, or by using a strap or chain wrench, if removed. A large worm drive hose clip can be used to clamp an L-shaped piece of steel strip to the clutch centre as a makeshift strap wrench.

8   With the nut and tab washer removed, it will be necessary to draw the clutch centre off the taper. This can be done using the Triumph service tool (No. D400) if this is available. Alternatively, two worm drive hose clips may be attached as shown in the accompanying photographs. These may then be used as a means of obtaining purchase with a conventional legged puller.

8 Removing the distributor

1   Two types of distributor clamp arrangement were used on the Terrier and early Tiger Cub models; All Terrier machines and Tiger Cubs up to engine number 57616 had an external clamp at the base of the unit. Subsequent models used an internal clamp operated by a screw which passed through the timing cover, just below the top edge. In either case, the distributor can be released after the retaining screw has been slackened.

7.2 Remove outer casing complete with stator assembly

7.4a Rotor is retained by large sleeve nut

7.4b Light off spacers and Woodruff key

7.6a Remove the screws, springs, spring cups and pressure plate

7.6b Clutch plates can now be displaced and removed

7.6c Lift away the clutch drum and primary drive

7.7 Clutch centre is retained by nut and tabwasher

7.8 Hose clips can be employed to facilitate removal

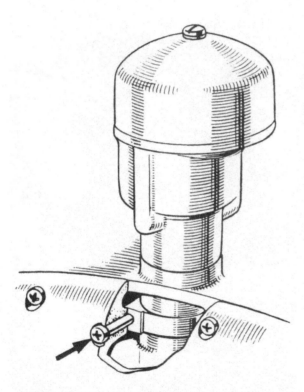
Fig. 1.1 Location of later type distributor clamp screw

9 Removing the contact breaker plate and automatic timing unit (September 1963 onwards)

1   Later model Tiger Cubs have the contact breaker and ATU incorporated in the timing cover. These must be removed before the timing cover can be released. Start by removing the inspection cover, which is retained by two screws. The contact breaker plate is held by two extended nuts. When these have been removed, the assembly can be pulled clear of the housing, and left suspended on the low tension leads.
2   The ATU is fitted onto a taper, and it is necessary to use either Triumph service tool D485, or to fabricate a similar tool, to effect its removal. After the retaining bolt has been removed,

two different threads will be visible, one of which is in the camshaft end, (the smaller, inner thread), the other being in the ATU boss. Obtain a bolt of similar dimensions to those of the retaining bolt, then cut it off so that when screwed into the camshaft thread, the top is about $\frac{3}{8} - \frac{1}{2}$ in from the end of the ATU. Use a hacksaw to cut a screwdriver slot in the end of the special bolt, then screw it into position. Obtain a bolt which will screw easily into the larger, outer thread. If this is tightened so that it bears upon the smaller special bolt, it will draw the unit off its taper.
3   The kickstart lever should now be removed, after slackening the securing nut on the end of the cotter pin threads, then striking the pin end to free it, or by slackening the pinch bolt, if this fitting is employed. Remove the gearchange pedal, which is secured by a clamp bolt arrangement. If the unit is still in the frame, disconnect the contact breaker leads at the rear of the unit, then remove the nine screws which retain the outer cover, and lift this away, guiding the leads through the hole in the inner casing.

10 Removing the inner timing cover and gearbox components

1   The inner half of the timing chest doubles as the gearbox end cover. If attention to the gearbox only is required, this can be done with the unit in place, there being no need to separate the crankcase halves as is current practice with unit construction engines.
2   Remove the felt seal which fits over the gearchange spindle, then disengage and remove the kickstart return spring and guide plate. Remove the two screws which retain the rectangular pressed steel cover, and lift this off to expose the camplate pivot. Extract the split pin which retains the complete boss to the pivot pin. The pin can now be withdrawn, leaving the camplate in position in the gearbox.
3   If the operation is being performed with the engine unit in the frame, remove the gearbox drain plug assembly, catching the oil content (about $\frac{1}{3}$ pint) in a suitable drain tray. The clutch cable should also be freed, after slackening it off at the handlebar lever, and then pulling it out of the back of the casing. Slacken and remove the six retaining screws, then carefully draw the cover away, taking care that the camshaft remains in position in the crankcase. Note that if the engine is still in the frame, and the camshaft does become displaced, the pushrods will drop clear of the rockers. It would then be necessary to remove the cylinder head to reassemble the valve train correctly.

4   The gearchange spindle assembly can be disengaged from the casing and removed after the securing bolt has been released. On some models, a gear position indicator is fitted, and this must be detached from the quadrant. If this is of the type which displays the gear position on the headlamp nacelle, it takes the form of a Bowden cable which is attached to a trunnion on the quadrant. A second type of indicator consists of a thread rod attached to the camplate, onto which is threaded an indicator blade which projects through a boss in the crankcase.

5   The kickstart spindle can be either left in position, or pulled out of its recess, taking care that the ratchet pawl and spring do not fly out, followed by the large thrust washer which sits in the recessed layshaft 1st (bottom) gear pinion. The gear clusters are now free to be pulled out of the casing, together with the two selector forks. The selector fork shaft will remain in position. The mainshaft top (4th) gear pinion takes the form of a sleeve gear to which the gearbox sprocket is bolted. This will remain in the casing, and need not be disturbed unless attention to it or the sprocket is required.

6   Whilst it is theoretically possible to slacken the gearbox sprocket nut to allow the sleeve gear to be pulled inwards and clear of the sprocket, it will be found that the nut is of very thin section, requiring the use of a correspondingly thin C spanner to effect its removal. It is thus easier to remove and separate the crankcase halves to give better access to the sprocket nut.

9.1 The contact breaker base plate assembly is held by two long nuts

9.2 Home-made extractor for ATU

9.3 Remove the right-hand outer cover, freeing cable

10.2a Release kickstart spring and backplate

10.2b Remove split pin and displace pivot pin to release quadrant

10.3 The inner cover can now be lifted away

10.4a Remove shouldered anchor bolt, and withdraw gearchange spindle

10.4b Disconnect gear position indicator and withdraw quadrant

10.5a Gear clusters and forks can now be lifted out of case ...

10.5b ... leaving the sleeve gear in position

11 Removing the camshaft, tappets and oil pump

1   If the engine unit has been removed and is being stripped prior to crankcase separation, all the above components may be removed. It should be noted, however, that there is no need to remove the oil pump and its drive unless specific attention is required. In the case of the above components requiring attention with the engine in place, bear in mind that the cylinder head must be removed to allow the pushrods and tappets to be relocated after the camshaft has been removed.

2   The camshaft and pinion can be lifted out of the casing and placed to one side, allowing the tappets to be displaced downwards. These should preferably be marked with tape, so that they can be refitted in their original positions. The plunger oil pump assembly can be removed after the two securing bolts have been released. Take care that the two $\frac{5}{32}$ in steel balls and the two springs which will be exposed when the pump body is lifted away, are not lost. Place all the pump components in a small container, to avoid loss.

12 Removing the crankshaft pinion assembly

1   The camshaft, or timing, pinion, was integral with the oil pump/distributor skew gear, on early models. This type of

arrangement involved a conical boss on the pinion assembly engaging in an internal taper in the mainshaft end, the assembly being retained by a long centre bolt. To remove the unit, the crankshaft must be locked by passing a bar through the connecting rod eye, the ends being supported on a wooden block at each side of the crankcase mouth. The securing bolt may now be slackened, and should be unscrewed by about one turn. If the correct Triumph extractor, part number D 398, is available, this may be screwed into the pinion assembly and the unit drawn off. Remove the extractor, unscrew the securing bolt completely, and lift the assembly away.

2   In the absence of the correct tool, a certain amount of ingenuity must be applied to devise a suitable alternative. It may be possible to modify a suitably-threaded scrap part to fit the internal thread of the boss, tapping the centre to accept an extractor bolt. Alternatively, a small legged puller may be used, provided there is enough clearance to fit it. Whatever the method used, care must be taken to avoid damage to the pinions or casing. If necessary, take the unit to a motorcycle repair specialist, who should have the equipment necessary to remove it safely.

3   Later models, fitted with a journal ball main bearing, use a different arrangement in which the skew gear and timing pinion are separate, sharing a common key on the mainshaft end. The two components can be slid off the shaft after the securing nut has been removed. Should they prove stubborn, a block of wood can be laid across the pinions, and tapped with a hammer to jar them loose.

---

13 Crankcase types: identification and construction

A   'One piece' type

The first type of crankcase was used on all Terrier models, and Tiger Cub models up to engine number 57616 in September 1959. The crank chamber and gearbox casing were formed almost exclusively by what is normally regarded as the right-hand crankcase half. The inner half of the primary chaincase doubles as a circular end cover for the crank chamber. The construction of the right-hand cover, or timing case, is very similar to that used on later models. Apart from the obvious lack of a join across the axis of the crankcase mouth, an easy means of identification is the external distributor clamp. This is worth knowing if second hand spares are being sought, as the early type of right-hand outer cover does not have a hole for a clamping screw.

B   Early two-piece type

From September 1959 to September 1963 (engine Nos from 57617 to 94599) a conventional two-piece crankcase was employed. In this type, the join passes directly across the centre of the crankcase mouth, the crank chamber being in two halves. Note that the distributor clamp on these models was re-located inside the timing case, the clamping screw passing through the outer cover, just below the top edge.

C.   Late two-piece type

From September 1963/engine number 94600 on, a modified two-piece crankcase was employed. The plain timing side bearing was dropped in favour of a journal ball type as used on the drive side. The distributor was omitted from this design, and a round blanking plug fitted in its place. The outer timing cover will be seen to have been modified to accept the contact breaker assembly and ATU.

General

It will be seen from the foregoing that although outwardly similar, there are three distinct types of engine casings employed. As they are not generally interchangeable, the differences have been outlined so that replacement parts can be easily identified. In the case of the two-piece design, it should be noted that it is particularly important to obtain the crankcase halves as a pair, as they are matched and jig-bored together. There is a danger that the main bearing bosses will not be correctly aligned if two odd halves are used, and this must be avoided as far as possible. If it is absolutely necessary to use unmatched halves, they must be checked for alignment before assembly.

11.2a Lift out camshaft and pinion as a unit ...

11.2b ... to allow tappets to be displaced and withdrawn

11.2c The oil pump is retained by two bolts

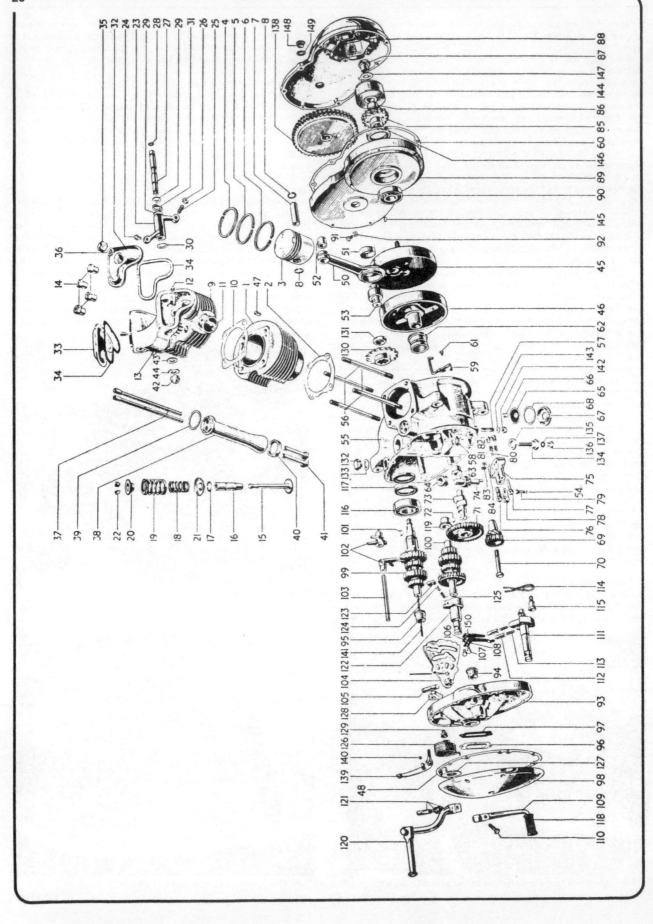

Fig. 1.2 Engine and gearbox unit – component parts – early type

1 Cylinder barrel
2 Cylinder base gasket
3 Piston
4 1st compression ring
5 2nd compression ring
6 Oil control ring
7 Gudgeon pin
8 Circlip – 2 off
9 Cylinder head
10 Cylinder head gasket
11 Exhaust pipe adaptor
12 Rocker cover stud – 2 off
13 Carburettor stud – 2 off
14 Nut – 4 off
15 Valve – 2 off
16 Valve guide – 2 off
17 Circlip – 2 off
18 Valve inner spring – 2 off
19 Valve outer spring – 2 off
20 Valve spring top collar – 2 off
21 Valve spring bottom collar – 2 off
22 Valve split cotter – 4 off
23 Valve rocker – 2 off
24 Ball ended pin – 2 off
25 Rocker adjuster – 2 off
26 Lock nut – 2 off
27 Rocker spindle – 2 off
28 Sealing ring
29 Thrust washer – 2 off
30 Thrust washer – 2 off
31 Spring washer – 2 off
32 Rocker inspection cover
33 Rocker inspection cover
34 Cover gasket – 2 off
35 Nut
36 Washer
37 Push rod – 2 off
38 Push rod tube

39 Washer
40 Washer
41 Tappet – 2 off
42 Domed nut
43 Washer
44 Washer
45 Flywheel assembly – LH
46 Flywheel assembly – RH
47 Dowel
48 Gasket
50 Connecting rod
51 Big end bush
52 Small end bush
53 Crankpin
54 Circlip
55 Crankcase assembly
56 Stud – 4 off
57 Stud
58 Dowel
59 Oil return pipe
60 Gasket
61 Bolt
62 Bearing – RH
63 Camshaft inner bush
64 Bush
65 Crankcase filter
66 Spring
67 Filter cap
68 Washer
69 Timing and distributor timing pinion
70 Bolt
71 Camshaft pinion
72 Woodruff key
73 Inlet and exhaust camshaft
74 Oil pump and distributor drive pinion
75 Oil pump body
76 Oil scavenge plunger
77 Oil feed plunger

78 Oil pump drive rod
79 Oil pump drive pin
80 Oil pump valve ball
81 Spring – 2 off
82 Plug – 2 off
83 Auxiliary spring – 2 off
84 Auxiliary ball – 2 off
85 Engine sprocket
86 Spacer
87 Nut
88 Inner cover – LH
89 Outer cover – LH
90 Main ball bearing – LH
91 Oil level plug
92 Washer
93 Inner cover – RH
94 Bush
95 Gearbox mainshaft bush
96 Inner cover plate
97 Inspection plate washer
98 Outer cover – RH
99 Mainshaft gear cluster
100 Layshaft gear cluster
101 Woodruff key
102 Selector fork – 2 off
103 Selector fork spindle
104 Cam plate
105 Cam plate spindle
106 Cam plate locating spring
107 Spring screw
108 Lockplate screw
109 Gearchange lever
110 Bolt
111 Gearchange, Spindle/Quadrant
112 Gear selector plunger
113 Spring
114 Return spring

115 Anchor pin
116 Bearing
117 Oil seal – 2 off
118 Rubber
119 Bush
120 Kickstart pedal
121 Cotter pin
122 Kickstart spindle
123 Kickstart pawl
124 Plunger
125 Plunger spring
126 Return spring
127 Kickstart return plate
128 Kickstart stop plate
129 Stop plate screw
130 Gearbox sprocket
131 Nut
132 Plug
133 Washer
134 Drain plug
135 Washer
136 Level plug
137 Washer
138 Clutch sprocket
139 Clutch operating lever
140 Ball
141 Clutch operating rod
142 Nut
143 Washer
144 Alternator rotor
145 Woodruff key
146 Oil seal
147 Tab washer
148 Plug
149 Washer
150 Cam plate locating spring

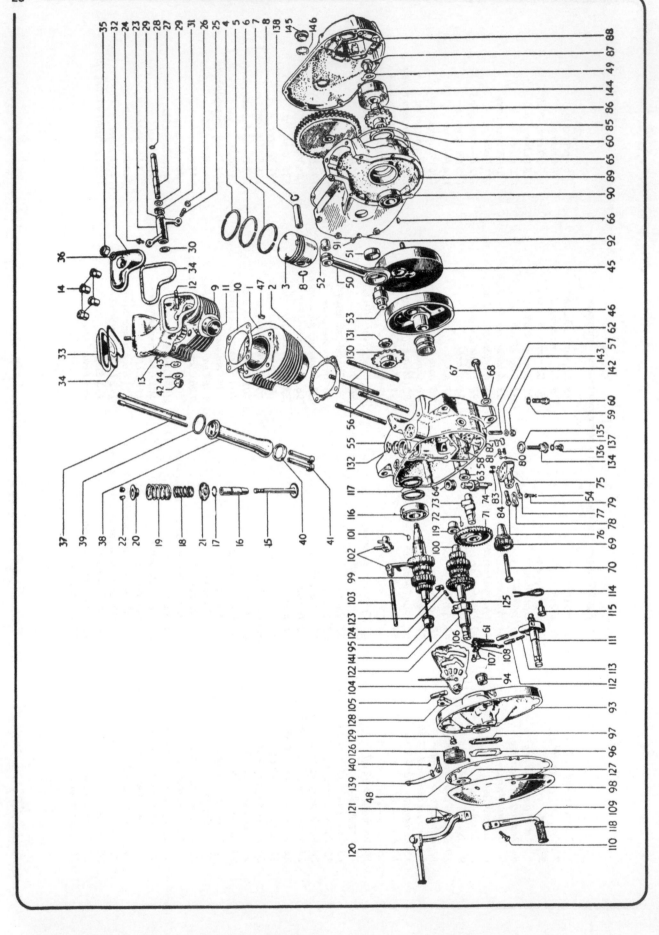

Fig. 1.3 Engine and gearbox unit – component parts – later pattern

1 Cylinder barrel
2 Cylinder base gasket
3 Piston
4 1st compression ring
5 2nd compression ring
6 Oil control ring
7 Gudgeon pin
8 Circlip – 2 off
9 Cylinder head
10 Cylinder head gasket
11 Exhaust pipe adaptor
12 Rocker cover stud – 2 off
13 Carburettor stud – 2 off
14 Nut – 4 off
15 Valve – 2 off
16 Valve guide – 2 off
17 Circlip – 2 off
18 Valve inner spring – 2 off
19 Valve outer spring – 2 off
20 Top collar – 2 off
21 Bottom cup – 2 off
22 Split cotter – 4 off
23 Valve rocker – 2 off
24 Ball ended pin – 2 off
25 Rocker adjuster – 2 off
26 Lock nut
27 Rocker spindle – 2 off
28 Sealing ring
29 Thrust washer – 2 off
30 Thrust washer – 2 off
31 Spring washer – 2 off
32 Rocker cover
33 Rocker cover
34 Cover gasket – 2 off
35 Retaining nut
36 Copper washer
37 Push rod – 2 off
38 Push rod tube

39 Washer
40 Washer
41 Tappet – 2 off
42 Domed nut
43 Copper washer
44 Copper washer
45 Flywheel – LH
46 Flywheel – RH
47 Dowel
48 Gasket
49 Tab washer
50 Connecting rod
51 Big end bush
52 Small end bush
53 Crankpin
54 Circlip
55 Crankcase – right
56 Stud – 4 off
57 Stud
58 Inner cover dowel
59 Washer
60 Oil filter
61 Camplate index spring
62 Main bearing – plain
63 Bush
64 Bush
65 Oil seal
66 Woodruff key
67 Bolt
68 Washer
69 Timing pinion
70 Bolt
71 Camshaft pinion
72 Woodruff key
73 Camshaft
74 Distributor drive pinion
75 Oil pump body
76 Scavenge plunger

77 Feed plunger
78 Pump drive rod
79 Connecting pin
80 Pump valve ball
81 Spring
82 Plug – 2 off
83 Auxiliary valve spring – 2 off
84 Auxiliary valve ball – 2 off
85 Engine sprocket
86 Spacer
87 Nut
88 Outer cover – left
89 Crankcase – left
90 Main ball bearing
91 Plug
92 Washer
93 Inner cover – right
94 Bush
95 Bush
96 Inner cover plate
97 Gasket
98 Outer cover – right
99 Mainshaft gear cluster
100 Layshaft gear cluster
101 Clutch key
102 Selector fork – 2 off
103 Selector fork spindle
104 Gearchange camplate
105 Spindle
106 Camplate index spring
107 Screw – 2 off
108 Lockplate
109 Gearchange lever
110 Bolt
111 Gearchange quadrant
112 Gear selector plunger – 2 off

113 Spring – 2 off
114 Return spring
115 Anchor bolt
116 Bearing
117 Oil seal – 2 off
118 Rubber
119 Bush
120 Kickstart pedal
121 Kickstart cotter
122 Kickstart quadrant
123 Pawl
124 Plunger
125 Plunger spring
126 Return spring
127 Return plate
128 Kickstart stop plate
129 Screw
130 Gearbox sprocket
131 Nut
132 Filler plug
133 Washer
134 Drain plug
135 Washer
136 Plug
137 Washer
138 Clutch
139 Clutch actuating lever
140 Ball
141 Clutch operating rod
142 Nut
143 Washer
144 Alternator rotor
145 Filler plug
146 Washer
147 Chaincase gasket

12.3a Lock crankshaft and release securing nut, ...

12.3b ... then remove pinion and skew gear

14 Separating the crankcase halves – general note

1   The condition of both the main bearings and the big end bearing can be checked before the crankcase is separated, thus obviating any unnecessary dismantling work. Refer to Section 18 of this Chapter for details.

15 Separating one-piece crankcase halves

1   The inner half of the primary chaincase, which forms the end cover for the crank chamber, is retained by one cross-headed screw, situated at the nine o'clock position in relation to the crankshaft, and two hexagon-headed bolts at two o'clock and four o'clock. With these removed, the casing can be pulled off. To assist separation, tap around the joint with a hide mallet, taking care not to damage the castings.

2   The upper and lower chaincase screw holes are designed to accept Triumph Z 129 extractors, in the absence of which, two screws of the correct thread can be modified by grinding a plain shank at the ends. When tightened, these will serve to draw the inner chaincase off the crankcase.

3   Once the inner chaincase/end cover has been removed, turn the crankshaft to BDC, so that the connecting rod is at its lowest point. The complete assembly can then be lifted out of the crank chamber, taking care not to damage the plain timing side bearing as the mainshaft is pulled through. The crankshaft assembly and the casings are now ready for cleaning and examination.

16 Separating two-piece crankcase halves

1   Slacken and remove the nine hexagon-headed bolts which secure the two crankcase halves. Using a hide mallet, tap around the joint to help break the seal. The drive side casing can now be drawn off, after which the crankshaft assembly may be pulled free of the timing side crankcase half. Note that the procedure is the same, irrespective of whether a journal ball or plain bush type of timing side main bearing is employed.

17 Examination and renovation: general

1   Before examining the parts of the dismantled engine unit for wear, it is essential that they should be cleaned thoroughly. Use a paraffin/petrol mix to remove all traces of old oil and sludge that may have accumulated within the engine.

2   Examine the crankcase castings for cracks or other signs of

damage. If a crack is discovered, it will require professional repair.

3   Carefully examine each part to determine the extent of wear, checking with the tolerance figures listed in the Specifications section of this Chapter. If there is any question of doubt, play safe and renew.

4   Use a clean, lint-free rag for cleaning and drying the various components. This will obviate the risk of small particles obstructing the internal oilways, causing the lubrication system to fail.

18 Big end and main bearings: examination and renovation

1   As mentioned in Section 14, preliminary examination of the big end and main bearings can take place before the crankcase halves are separated. The big end is checked for wear by arranging the connecting rod in its TDC position. Grasp the connecting rod, and pull it firmly upward and then downwards, making sure that any end float, which is intentional, is not mistaken for big end play. Any discernible up-and-down movement will indicate that the bearing is worn and in need of renewal.

2   Main bearing wear can be checked by grasping each mainshaft in turn and attempting to move it in relation to the bearing. This check should be undertaken with some degree of vigour, as oil in the bearing(s) will tend to mask any movement. Any discernible play is indicative of the need for renewal, and in any case the bearings should be washed out and closely examined after parting the crankcase halves.

3   Wash the bearings carefully in clean petrol to remove any residual oil, then check for free play, and spin them to check for any sign of roughness or grittiness. The journal ball type main bearings can be tapped out of the cases, after heating the cases to about 100°C in an oven. If an oven is not available, boiling water can be used to good effect. If the cases are heated as described above, the bearing will be released fairly easily. New bearings should be fitted in a similar manner, ensuring that they seat correctly in the crankcase bores.

4   Where applicable, examine the plain timing side main bearing for wear and scuffing, and check that the crankshaft fits smoothly in place with no discernible free play. It should be noted that the bush **must** be in good condition or pressure in the oil feed to the big end bearing will be lost. The bush is renewable, but must be line-reamed after fitting. As this is beyond the scope of most owners, it is recommended that the crankcases are entrusted to a reputable repair specialist or engineering works for this job to be done.

5   If the big end bearing requires attention, it is again recommended that the work is entrusted to a specialist who will have

the facilities for pressing apart the flywheels, reassemble them, and re-true them, the latter stage being most important if a long and vibration-free life is to be expected. Renewal of the big-end bearing requires some experience and the use of various measuring instruments to ensure the requisite accuracy in re-aligning the flywheels. It is therefore recommended that the complete flywheel assembly be entrusted to a competent motorcycle repair specialist for the work to be carried out. The rebuilt assembly must be realigned with a lathe.

6   If it is absolutely necessary that the flywheel assembly be dismantled at home, proceed as follows: Obtain Triumph tool No Z101, or fabricate a similar tool from a piece of steel bar. Drill two locating holes to match the threaded hole centres in the flywheel. Two $\frac{3}{8}$ in bolts (high tensile steel) will be required to fit through these holes. Drill and tap the centre of the tool to accept a suitable high tensile bolt, the end of which should be ground to a blunt taper.

7   The accompanying drawings show the tool assembled to remove firstly the left-hand flywheel, and then the crankpin. Before separation starts, however, obtain an engineering set square, and scribe a line across the two flywheel edges to aid truing when reassembly takes place. Once dismantled, the worn parts should be renewed. Note that it is always advisable to renew the bearing and connecting rod as a matched assembly, as the latter may well be slightly oval, if not obviously worn or damaged.

8   New big end bearings are usually available as an assembly, complete with an exchange connecting rod, in which case the complete unit will be ready for refitting. In other cases the bearing is supplied separately. In the latter instance the outer race of the old bearing will require pressing from the connecting rod and the new outer race pressed in. The new race will probably require honing so that the rollers are a perfect fit.

9   Provided that the roller track of the crankpin and that of the outer race is not scored or pitted, it is permissible to take up a small amount of wear by honing the outer race and fitting slightly oversize rollers. This method of bearing reclamation is only recommended as a last resort and should not be applied to a competition machine where the stress is consistently greater than that encountered by a road going machine.

10   When realigning the flywheel assembly the maximum permissible run-out is 0·002 in (0.05 mm) at the mainshaft ends.

11   During reassembly, take great care that the oilways and sludge traps are cleaned out properly, and that the crankpin is fitted so that the oilways align correctly. This should be checked by forcing oil through them with a pressure type oil can. To clean out the sludge trap remove the plug situated in the rim of the right-hand flywheel. Tighten the plug securely on refitting and stake it in place to prevent it from slackening.

12   A worn big end bearing can be recognised by a characteristic knock and accompanying vibration. Do not use the machine in this condition since there is grave risk of a broken connecting rod and accompanying severe engine damage, especially if the rod fractures whilst the engine is highly stressed.

18.1a Check for up and down movement in big-end bearing ...

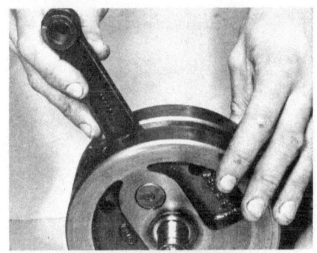

18.1b ... taking care not to confuse it with sideplay

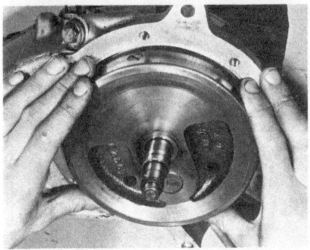

18.2 Any discernible movement in main bearing will require renewal

18.3 Bearings may be drifted from case, preferably after heating

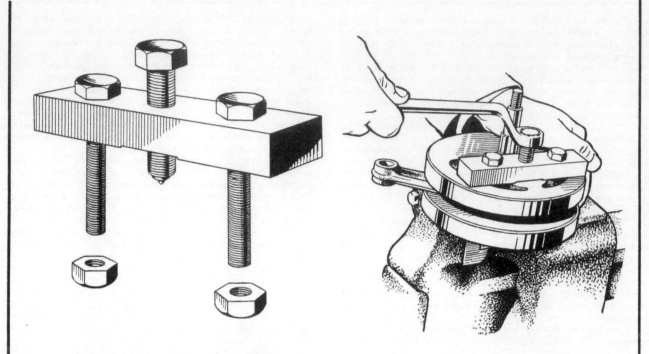

Fig. 1.4 Flywheel separating and assembly tool

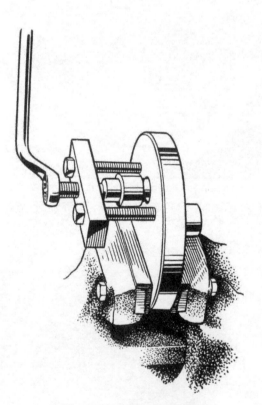

Fig. 1.5 Removing the drive side flywheel

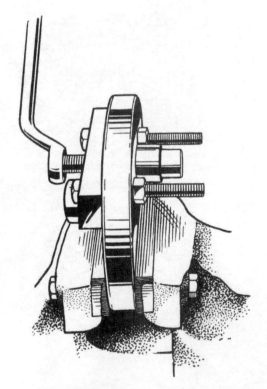

Fig. 1.6 Removing the crankpin

Fig. 1.7 Refitting the crankpin

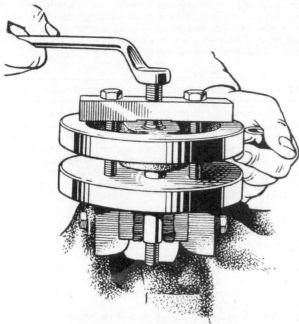

Fig. 1.8 Reassembling the flywheel halves

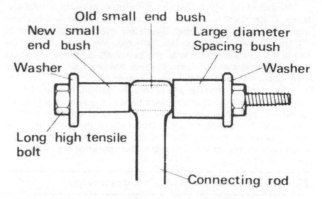

Fig. 1.9 Using drawbolt method to renew small-end bush

19 Small end bush: examination and renovation

1   The small end bearing of the connecting rod takes the form of a phosphor bronze bush with an oil hole to provide lubrication for the gudgeon pin and inner bush surfaces. The gudgeon pin should be a good sliding fit in the bush, without evidence of any vertical play. When play develops, the bush must be removed and renewed, but it is first advisable to check whether the wear has not occurred in the gudgeon pin itself.

2   A simple extractor can be made to remove the old bush and at the same time pull the new bush into location. Refer to the accompanying diagram. It must be aligned so that the oil hole registers exactly with the slot in the top of the connecting rod, or the bearing will be starved of oil. When in the correct position, it must be reamed out so that the gudgeon pin is once again a good sliding fit, with a complete absence of vertical play.

20 Timing pinion and camshaft: examination and renovation

1   It is unlikely that the timing pinion will require attention unless the machine has covered an exceptionally high mileage, or if the entry of some foreign body has caused accidental damage.

2   Check that the camshaft is a good fit in its supporting bushes, with no discernible free play. If worn, these bushes can be driven out of the casings after the cases have been heated to about 100°C in an oven, or by pouring boiling water over the casing.

3   After fitting new bushes in a similar manner, check that the camshaft will turn easily. If there is any sign of tightness, it will be necessary to line-ream the bushes. This is best done by using an expanding reamer, taking out a small amount of metal at a time until the shaft end is a light sliding fit.

4   An alternative method of renewal is to employ a drawbolt arrangement such as that described for small end renewal. It should be noted, however, that the new bush cannot be used to draw the old bush out of the casing, as the bushes are headed.

5   The camshaft should be checked carefully for signs of wear or damage, paying particular attention to the lobe surfaces and the bearing journals. If these are worn or scored, it is advisable to try to obtain a new camshaft, although it is possible to have a worn component reclaimed should this prove necessary. A number of engineering firms offer this service, and advertise regularly in the motorcycle press.

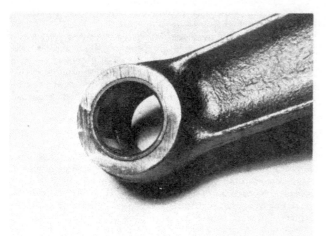

19.1 Inspect small end bush for wear in conjunction with gudgeon pin

19.2 If new bush is fitted, ensure that oil slot is aligned

6   It should be noted that the cam lobes have hardened surfaces. Once these have worn through, cam wear will become more pronounced and rapid, ultimately affecting the amount of lift and the valve timing. The cam followers must always be examined in conjunction with the camshaft, as these will show corresponding signs of wear. It is normal practice to renew the cam followers and camshaft as an assembly.

7   If it is found necessary to renew the camshaft, keep the old item as a pattern so that a replacement of the same type can be found. As mentioned previously, a number of different profiles were available during the production life of the machine, each giving different engine characteristics, and requiring various valve clearances, ignition timings and carburettor settings.

21 Cylinder barrel: examination and renovation

1   Unless the cylinder barrel is new or has recently been rebored, there will be a lip at the top of the bore that denotes the limit of travel of the top piston ring. The depth of this lip will give some indication of the extent to which wear has taken place, even though the amount of wear is not evenly distributed.

2   Check first that no obvious damage to the bore surface is evident. If this is the case, a rebore and a new piston will be necessary, and further examination will be pointless. If broken piston rings were encountered during dismantling, look carefully for signs of scoring on the bore surface, as small hairline grooves are often caused by the broken ends of the rings and are easily missed.

3   If the bore surface appears to be undamaged, the amount of wear should be assessed. This is best done using an internal micrometer, taking one measurement at the very top or bottom of the bore, to give the original bore size, and a second reading about $\frac{3}{4}$ in below the ridge at the top of the bore. If the first reading is subtracted from the second, the amount of wear will be apparent.

4   A more practicable method is to position one of the compression rings squarely in an unworn portion of the bore, and measure the end gap. Repeat the measurement at a worn part of the bore and subtract the first reading from the second. Divide the result by three to give an approximate indication of the amount of wear.

5   The normal wear limit is + 0·005 in (0·127 mm) and is indicative of the need for a rebore and an oversize piston. It will be found in practice that with some experience it is possible to establish whether or not a rebore is required by interpreting the various symptoms, i.e. piston slap, compression loss and the appearance of the bore and piston.

6   If the amount of wear is slight, it is acceptable to use the bore again, having first checked whether new rings are required. This can be done by inserting a ring in the bore and measuring the end gap after pressing it down squarely in the bore with the piston. If the end gap is in excess of 0·010 in (0·24 mm) or if there is any sign of wear on the ring faces, they must be renewed. Check the end gap of the new rings before fitting them to the piston. This must be between 0·008 and 0·010 in (0·2 – 0·24 mm). If necessary, the ring ends can be filed very carefully with a fine swiss file to obtain the correct clearance. Do not omit to file a small 45° chamfer across the ends to prevent scoring. Note that this measurement should be made with the ring fitted in an unworn portion of the bore.

7   If new rings or a new piston assembly are to be fitted to a partly worn bore, it is important to remove the polished surface finish of the cylinder wall to enable the new rings to bed themselves in. Ideally this should be done by honing or with one of the proprietary glaze-breaking attachments for power drills. In the absence of these facilities, the glaze may be removed with some careful work with fine emery cloth. The object is to *just* remove the shine from the surface, and care must be taken not to remove excessive amounts of metal. The bore should be wiped with a petrol soaked rag to remove any residual abrasive particles after the work is complete.

8   The outside of the cylinder barrel should be cleaned carefully, and any obstruction between the fins removed. Cast iron barrels benefit from the application of cylinder black, which is obtainable as an aerosol spray or as a brush paint. This has the effect of improving heat dissipation as well as enhancing the appearance of the cylinder barrel.

22 Piston and piston rings: examination and renovation

1   Attention to the piston and piston rings can be overlooked if if a rebore is necessary, since new replacements will be fitted.

2   If a rebore is not considered necessary, examine the piston carefully. If score marks are evident, or if the skirt is badly discoloured as the result of exhaust gases by-passing the rings, a new piston and rings should be fitted. Check that the correct size is fitted; if the original piston was oversize, the amount will be stamped on the crown.

3   Remove the carbon from the piston crown, using a blunt scraper which will not damage the surface. Clean away all carbon deposits from the valve cutaways and finish off with metal polish so that a clean, shining surface is achieved. Carbon will not adhere so readily to a polished surface.

4   Check that the gudgeon pin bosses are not worn or the circlip grooves damaged. Check that the piston ring grooves are not enlarged. Side float must not exceed the amount laid down in the Specifications.

5   When fitting new piston rings, or if the engine has seen long service, a check should be made to ensure there is no build up of carbon behind the rings, or in the grooves of the piston. Any build up should be removed by gentle scraping.

6   Do not omit to examine the gudgeon pin. Wear is most likely to occur where the pin passes through the small end bush and can be measured with a micrometer. If there is play at the small end bush and the bush itself is in good order, the gudgeon pin should be renewed.

23 Valves, valve springs and valve guides: examination and renovation

1   Before the valves, valve springs and valve guides can be examined, it is necessary to remove the valves from the cylinder head. This is accomplished by means of a valve spring compressor, which will compress each spring sufficiently to permit the split collets to be removed from each valve stem. When the collets have been detached, the compressor can be unscrewed until the valve spring tension is released. The valve, valve springs and collar can then be removed, leaving the valve guide in the cylinder head. To avoid confusion during reassembly, do not allow the valve and associated components to become mixed as some of the parts are different.

2   After cleaning the valves to remove all traces of carbon and burnt oil, examine the heads for signs of pitting and burning. Examine the valve seats in the cylinder head. The exhaust valve and its seat will probably require the most attention because it is the hotter running. If the pitting is slight, the marks can be removed by grinding the seat and the valve heads together, using fine valve grinding compound.

3   Valve grinding is a simple, if somewhat laborious task, carried out as follows. Smear a trace of fine valve grinding compound (carborundum paste) on the seat face and apply a suction grinding tool to the head of the valve. Oil the stem of the valve and insert it in the guide until it seats in the grinding compound. Using a semi-rotary motion, grind-in the valve head to its seat, using a backward and forward motion. It is advisable to lift the valve occasionally, to distribute the grinding compound more evenly. A light compression spring can be placed under the valve head to assist the lifting off process, especially if an oscillatory valve lapping tool is used for the grinding process. Repeat this application until an unbroken ring of light grey matt finish is obtained on both valve and seat. This denotes the grinding operation is now complete. Before passing to the other valve, make sure that all traces of the valve grinding compound

have been removed from both the valve and its seat and that none has entered the valve guide. If this precaution is not observed, rapid wear will take place due to the highly abrasive nature of the carborundum paste.

4   When deep pits are encountered, it will be necessary to use a valve refacing machine and a valve seat cutter, set to an angle of 45°. Never resort to excessive grinding because this will only pocket the valves in the head and lead to reduced engine efficiency. If there is any doubt about the condition of a valve, fit a new one.

5   Examine the condition of the valve collets and the groove on the valve stem in which they seat. If there are any signs of damage, new parts should be fitted. Check that the valve spring collar is not cracked. If the collets work loose or the collar splits whilst the engine is running, a valve could drop in and cause extensive damage.

6   Before grinding in the valves, the clearance between each valve stem and its guide should be checked. A good indication of worn valve guides is the presence of bluish smoke in the exhaust on over-run. This is caused by oil being drawn past the inlet valve stem by the depression in the inlet tract, and subsequently passing into the combustion chamber and burning.

7   Excessive play between a valve and guide will cause high oil consumption and oiling up of the cylinder, due to oil finding its way down the valve stems. Low compression may also be encountered due to tilting of the valve faces at the seat. A new guide may be fitted to an old valve and vice versa, provided that the old component being used is in good condition.

8   The valve guides are a tight drive fit in the cylinder head, and may be drifted from position, using a drift of suitable proportions. If at all possible, a double diameter drift should be used. There is no need to heat the cylinder head, although where an aluminium component is fitted, this will facilitate removal. Take care when removing cast-iron valve guides as they are prone to fracturing, thereby making final removal more difficult. On the exhaust valve guide it is worthwhile removing the build up of carbon from the outer diameter before trying to drift the guide from position.

9   When refitting new valve guides, ensure that the locating circlip is positioned correctly in its groove. The clip serves as a means of positive location for the guide, and prevents it from becoming displaced should it work lose during use. Drive the guide in carefully, with the surrounding area well supported. A double diameter drift should be used to prevent the edge of the guide from becoming burned or distorted. The guide will be correctly positioned when the circlip *just* contacts the cylinder head.

10 It is possible that a valve guide may work loose in the cylinder head, making the fitment of a standard replacement guide impracticable due to the bore in the cylinder head having become enlarged. The best course of action, short of obtaining another cylinder head, is to obtain a new guide of oversize external diameter. If necessary, most motor engineering companies will be able to make up a suitable guide to pattern or drawing, and bore the cylinder head to suit.

11 Whenever new valve guides are fitted, the valve seats must be re-cut with a 45° seat cutter, to ensure that the valve face is perfectly concentric with the seat. The special cutting tool is of the same type used for many car engines and therefore most garages are able to undertake the work.

12 Check the free length of the valve springs, which should be as follows:

Inner spring:	$1\frac{9}{16}$ in (40 mm)
Outer spring:	$1\frac{5}{8}$ in (41·3 mm)

If shorter than the above lengths, or if there is any doubt as to their condition, they should be renewed as a set. If new valves and/or guides have been fitted, it is worth renewing the springs as a matter of course, in view of their low cost. Note that heavyweight valve springs, identified by a dab of red paint, were fitted to T20S models from engine no 45312, T20SL, T20SS, T20SH and T20SC models. These springs were fitted in conjunction with a sports camshaft. Tired valve springs can impair performance to a marked degree by allowing valve bounce to occur, thus preventing the engine from attaining its designed maximum speed.

24 Cylinder head: examination and renovation

1   Remove all traces of carbon from the cylinder head and valve ports, using a soft scraper. Extreme care should be taken to ensure the combustion chamber and valve seats are not marked in any way, especially when an alloy head is being cleaned up. Finish by polishing the combustion chamber with metal polish so that carbon does not adhere so easily. Never use emery cloth since the abrasive particles will become embedded in the metal, especially in the case of alloy components.

2   Check to make sure the valve guides are free from oil or other foreign matter that may cause the valves to stick.

3   Make sure the cylinder head fins are not clogged with oil or road dirt, otherwise the engine may overheat. If necessary, use a wire brush.

4   Reassemble the cylinder head by replacing the valves after oiling their stems. Compress each set of valve springs in turn, making sure the split collets are located correctly before the compressor is released. A light tap on the end of each valve stem, after reassembly, will act as a double check.

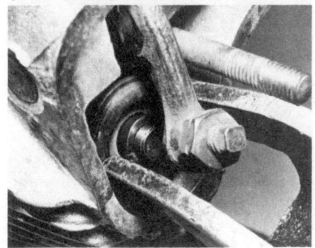

23.1a Compress valve spring to release collets

23.1b The valve can now be removed from cylinder head

25 Rockers, rocker spindles and rocker covers: examination and renovation

1   The rockers can be removed from the cylinder head, after the pivot pins have been driven out. Support the area around the pivot pin bosses, and drive them out from the oil feed side. As the pin is withdrawn, the rocker, one Thackeray spring washer and three plain thrust washers will be displaced. Clean the various components in petrol, then dry them off before further examination.

2   If the rocker shafts have worked loose in the cylinder head, it is likely that the bosses will have become oval. It is unwise to continue using the engine if this situation has arisen, as valve operation will become noisy and inaccurate, and the bosses will almost invariably leak oil. The cylinder head may be repaired, either by boring out and bushing the affected area, or by having the bosses built up by argon-arc welding, and then machined back to their original size.

3   Check the bore of the rockers and the bearing surface of their respective pivot shafts. If the fit between them proves to be sloppy, they should be renewed as a pair. The ball-ended pin at the end of the rocker should be examined for wear in conjunction with the relevant pushrod. Check each pushrod for straightness by rolling it on a surface plate or a sheet of glass. A bent pushrod should be renewed as a matter of course, as subsequent failure is likely if it is used again.

4   The light alloy rocker covers are unlikely to be damaged unless the central securing bolt has been overtightened at some stage. In the event that this has caused cracking around the boss, it may be repaired by argon-arc welding, or as a temporary expedient, by sealing the cracks with an epoxy adhesive. The obvious alternative to the above methods is to obtain new or second hand replacements.

5   When reassembling the rocker components, the following sequence should be adhered to:

Slide the pin in from the pushrod tunnel side of the cylinder head, and fit the thrust washer, followed by the rocker itself, pushing the pin through so that the end is flush with the rocker bore. Assemble a thrust washer on each side of the Thackeray spring washer, holding them together with a pair of pointed-nosed pliers. These components can then be squeezed together to allow them to be fitted between the end of the rocker and the inner face of the pivot shaft boss. The pivot shaft can now be driven into position, again using a soft drift.

6   Note that the external banjo unions, which connect the oil feed to each rocker shaft, are sealed by two copper washers each. A $\frac{3}{8}$ in copper washer fits between each union and the cylinder head, the outer items being $\frac{5}{16}$ in. Copper washers may be used indefinitely, provided that they do not become damaged or scored. It is usual to anneal copper washers prior to re-use. This allows the softened copper surface to conform with any irregularities, thus forming an effective seal. Copper is an unusual metal in that the process used to anneal it is much the same as that used to harden most other metals. The washer should be held in the flame of a gas ring or a blowlamp, and heated to a cherry-red colour. Remove the washer from the flame, and plunge it into cold water to quench it.

7   If scored or damaged, the rocker feed unions can be reclaimed, by lapping the sealing faces on fine emery paper fixed to a flat surface.

26 Gearbox components and kickstart assembly: examination and renovation

1   The gearbox consists of two clusters; the mainshaft, on which the gearbox sprocket and clutch are mounted, and the layshaft. The gearbox rarely gives trouble unless lubrication has been neglected at some stage.

2   Wash the components in petrol to remove all traces of oil, then allow the petrol to evaporate or blow dry with compressed air. Examine each of the gear pinions for chipped, worn or broken teeth. Reject any that have such defects and fit new components. Check for worn dogs on the ends of each pinion, especially those where the edges are rounded. Worn dogs are a frequent cause of jumping out of gear and renewal of the pinions concerned is the only effective remedy.

3   Gear selection faults are often caused by bent or worn selector forks or excessive side play in the forks as the result of seizure. Damage of this nature will be self-evident and replacement with new parts is essential.

4   Worn bearings are easy to renew, particularly those of the ball journal type. If a bearing of the ball journal type is suspect, it should be removed and washed out with a petrol/paraffin mix. If any signs of play are evident, or if the bearing runs roughly when it is turned, it should be rejected and renewed.

5   The kickstart assembly takes the form of a spindle with a spring loaded pawl, the latter operating on ratchet teeth cut on the inside of the layshaft first (1st) gear pinion. Remove the pawl from the spindle by displacing it sideways, taking care not to lose the plunger and spring, which will drop free.

6   Examine the end of the pawl in conjunction with the internal ratchet teeth. After extensive periods of use, these parts can be expected to wear, eventually leading to the ratchet slipping when attempting to start the engine. If worn, it will be necessary to renew the part(s) in question to effect a repair. Note that the pawl spring should also be renewed as it can cause the ratchet to slip if it has become weakened.

7   If it becomes necessary to renew the kickstart spindle, note that there are two types available, depending on the model. Most roadster Cubs were equipped with a cotter pin fixing kickstart lever, whilst the competition and sports models utilised a splined type. A new spindle must be selected accordingly.

8   It is worthwhile noting that the vast majority of kickstart spindle and lever renewals are necessary only because the pinch bolt or cotter pin has not been kept fully tightened. If even slightly loose, wear in these parts is greatly accelerated, and once slightly worn, wear will continue at an increasing rate despite any last-minute attempts to slow it.

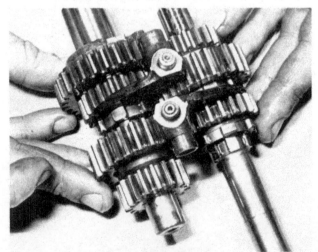

26.2 Gearbox components must be cleaned prior to inspection

26.3a Look for wear on face of quadrant ...

26.3b ... and on selector fork ends

26.5a Kickstart spindle incorporates ratchet mechanism

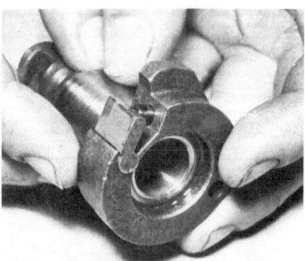
26.5b Pawl can be displaced sideways ...

26.5c ... to permit removal of plunger ...

26.5d ... and pawl spring, for inspection

26.5e Ratchet teeth must also be checked for wear

27 Clutch: examination and renovation

1   Check each clutch plate to make sure that it is completely flat and free from buckles. Reject any that have become distorted otherwise clutch troubles will persist.

2   The tongues at the edges of each inserted plate should be free from burrs or other damage. It is permissible to remove any burrs by dressing the edges of each tongue with a file. Make sure they remain square. Reject any plates where the tongues show signs of splitting or cracking. A repair is impracticable especially in view of the relatively cheap cost of replacements.

3   Check also the tongues at the centre of each plain plate. Burrs can be dressed in a similar manner. Make sure the surface of each plate is smooth and free from blemishes.

4   The linings of each inserted plate should be inspected. The amount of wear that has taken place can be checked against the Specifications at the beginning of this Chapter.

5   Unless the lining material is scored, charred or worn through to the backing metal, it is permissible to re-use the plates. If clutch slip has been a problem, it will obviously be necessary to renew the friction plates and clutch springs, as a matter of course.

6   Check the condition of the clutch springs. The free length must be between 1·66 in (42·1 mm) and 1·53 in (38·9 mm). If the springs have taken a permanent set below this figure, they should be renewed as a set, otherwise clutch slip is likely to occur.

7   The clutch pressure plate can be treated in a similar manner to the plain plates, ensuring that no discernible warpage is evident. The outer drum and chainwheel assembly should be checked for signs of burring or other damage, particularly in the region of the clutch plate slots.

8   After an extensive period of service, the tongues of the clutch plates tend to make indentations in the edge of each slot, which eventually prevent the clutch from freeing completely. The edges should be dressed with a file until they are square and completely free from indentations. Provided too much metal has not been removed, the additional amount of backlash in the plates will not be of too great consequence.

9   The outer drum runs on a caged ball race. It is not normally subject to excessive wear, but the amount of play should be checked, and the bearing renewed if worn. Note that the balls will be displaced if the split bronze race is removed, so care must be taken to avoid loss. When reassembling the bearing, the balls can be stuck in place on the race with dabs of grease. It is permissible to renew the sixteen $\frac{5}{32}$ in (0·156 in, 3·96 mm) balls if the other bearing surfaces are unmarked.

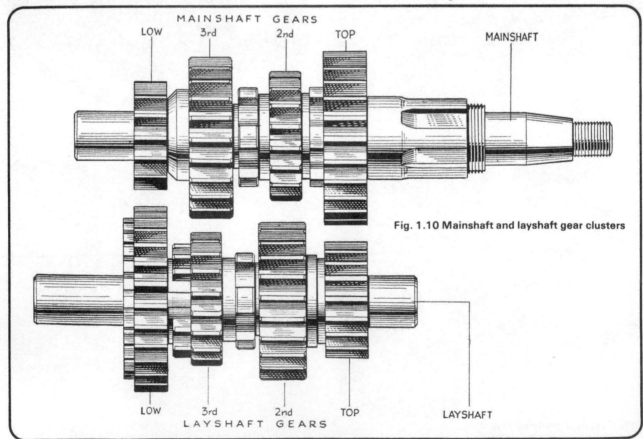

MAINSHAFT GEARS

LOW   3rd   2nd   TOP   MAINSHAFT

Fig. 1.10 Mainshaft and layshaft gear clusters

LOW   3rd   2nd   TOP   LAYSHAFT

LAYSHAFT GEARS

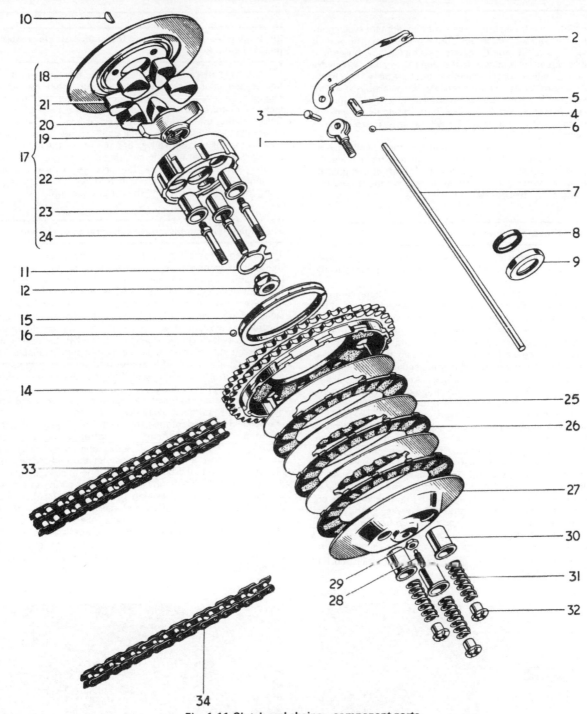

Fig. 1.11 Clutch and chains – component parts

1	Clutch operating lever assembly	13	Complete clutch assembly	24	Pin – 3 off
2	Lever	14	Sprocket and housing	25	Driven plate – 3 off
3	Pin	15	Bearing ring	26	Friction plate – 3 off
4	Cup	16	$\frac{5}{32}$ in Steel ball – 16 off	27	Pressure plate
5	Split pin	17	Shock absorber – complete	28	Adjuster
6	Steel ball	18	Back plate	29	Nut
7	Clutch operating rod	19	Spider	30	Cup – 3 off
8	Oil seal	20	Drive rubber – 3 off	31	Spring – 3 off
9	Oil seal retainer	21	Rebound rubber – 3 off	32	Nut – 3 off
10	Woodruff key	22	Centre	33	Primary chain
11	Tab washer	23	Cup – 3 off	34	Rear chain
12	Nut				

10 The clutch centre incorporates a rubber segment type shock absorber. This rarely gives trouble, but if it does become worn it is advisable to renew the complete assembly. It will be found that the backplate is retained by the three stud ends, which are staked into position during assembly, the backplate then being faced off square. Unless suitable facilities are available, it will be difficult to overhaul the shock absorber unit, even though the three large and three small rubber inserts should be obtainable. Probably the best course of action would be to try to obtain a good second-hand unit.

11 The clutch pushrod should be checked for straightness by rolling it on a sheet of plate glass or on a surface plate. Its bronze supporting bush in the hollow mainshaft end is unlikely to suffer much wear, but it may be drilled out and a new one pressed into place, if necessary.

28 Engine reassembly: general

1 Before reassembly is commenced, the various engine components should be thoroughly cleaned and placed close to the working area.

2 Make sure all traces of the old gaskets have been removed and the mating surfaces are clean and undamaged. One of the best ways to remove old gasket cement is to apply a rag soaked in methylated spirit. This acts as a solvent and will ensure the cement is removed without resort to scraping, with the consequent risk of damage. More stubborn deposits may be removed using a soft brass wire brush of the type used for cleaning suede shoes, to avoid scratching the surfaces. Always use gaskets of the correct material, and where considered necessary, apply a suitable gasket compound to one or both mating surfaces.

3 Gather together all the necessary tools and have available an oil can filled with clean engine oil. Make sure all the new gaskets and oil seals are to hand; nothing is more infuriating than having to stop in the middle of a reassembly sequence because a vital gasket or replacement part has been overlooked.

4 Make sure the reassembly area is clean and that there is adequate working space. Refer to the torque and clearance settings wherever they are given; many of the smaller bolts are easily sheared if they are over-tightened. Always use the correct size spanner and screwdriver, never a wrench or grip as a substitute. If some of the nuts and bolts that have to be replaced were damaged during the dismantling operation, it is advisable to renew them. This will make any subsequent reassembly and dismantling much easier.

29 Engine reassembly: fitting the mainshaft sleeve gear and gearbox sprocket

1 Lubricate the mainshaft sleeve gear, and slide it into position in the gear casing. It will be noted that the sprocket engages on splines on the sleeve gear, and that the nut is locked by centre punching it at one of the spline indentations. Before sliding the sprocket into position, mark the sleeve gear indentation positions with radial chalk marks on the sprocket face, as the splines themselves are masked by the nut.

2 Fit the sprocket and the securing nut, using a chain wrench or the drive chain to lock the sprocket whilst the nut is tightened. Use a sharp centre punch to peen the nut into place, using the chalk marks to find the correct position.

30 Engine reassembly: joining the crankcase halves

1 Joining the crankcase halves is a straightforward reversal of the separation operation described in Sections 15 and 16 of this Chapter. The crankshaft assembly should be positioned in the right-hand crankcase half, irrespective of the type of casing employed. No gasket is fitted between the two faces, but a smear of non-hardening gasket cement should be applied to ensure that the joint is air and oil tight. If the condition of the mating faces is in any way suspect, it is advisable to use one of the silicone rubber 'plastic gasket' compounds available. These compounds are designed to conform to any small irregularities, and will form an excellent seal if used in accordance with the manufacturer's directions.

2 Check the condition of the gearbox mainshaft seal in the back of the primary chaincase section of the left-hand crankcase half. If damaged, it will be necessary to renew it together with its retainer, the latter being peenedinto position.Check that the bronze sleeve gear bush is not rough or sharp at the end. If necessary, chamfer the end of the bush to prevent damage to the oil seal when fitting the casing half.

3 Assemble the left-hand casing half, easing the seal over the mainshaft end to avoid tearing the seal face. Fit the securing bolts, nine, in the case of the two-piece crankcase, or three on earlier models. Tighten the screws diagonally and evenly, checking that the crankshaft still turns easily. Note that if the crankshaft appears stiff, it indicates that there is some mal-alignment. This is unlikely, but must be investigated and rectified, if evident.

29.1a Sprocket on right is very badly worn – compare with new component (left)

29.1b Gearbox sleeve gear runs in journal ball bearing

41

29.1c Mark spline positions with chalk before ...

29.1d ... tightening securing nut. Nut must be punched to lock it in position (see main text)

30.1 Liberally oil all bearing surfaces, then fit crankshaft

30.3 Fit left-hand casing half, taking care not to damage seal

31 Engine reassembly: fitting the timing pinion, tappets, camshaft and oil pump

A. One-piece crankcase models
1 Set the crankshaft in the TDC position, and arrange the distributor shaft so that the driving slot is parallel to the crankshaft axis. Slide the timing pinion into position, twisting it to engage with the distributor shaft skew gear.
2 Before fitting the retaining bolt, measure its length to the underside of the head. This must not exceed 1·59 in (1 $\frac{19}{32}$ in, 40·48 mm). If the bolt is longer than this there is a risk of its obstructing the oil feed to the big end bearing. If necessary, file the bolt to size or obtain a new replacement item. Tighten the securing bolt, holding the connecting rod to prevent crankshaft rotation. Take care not to overtighten the bolt, as the wedge action of the taper will tend to spread the mainshaft journal.

B. Two-piece crankcase models
3 Set the crankshaft in the TDC position, and fit the long Woodruff key to the mainshaft. On models fitted with a distributor, arrange the driving shaft so that the slot is parallel with the crankshaft axis. Slide on the bronze skew gear, turning the crankshaft slightly to ensure that it meshes correctly.
4 Slide the timing pinion onto the mainshaft end with the flange innermost, then fit the securing nut. Lock the crankshaft

by passing a bar through the small end eye, or by selecting top gear and applying the rear brake if the engine is in the frame, then tighten the securing nut.

C. All models
5 Lubricate and insert the two tappets into their respective bores, then offer up the camshaft. It will be noticed that there is a timing mark between two of the camshaft pinion teeth. This must align with a corresponding mark on one of the timing pinion teeth. After fitting the camshaft, turn the engine over once or twice, then check that the marks are correctly aligned.
6 On models fitted with an internal distributor clamp, this should be positioned in the slot in the casing. If the oil pump has been removed for examination, it should now be refitted. It is important that the mating faces of the pump body and the casing are wiped clean, and a new gasket used. Note that the two springs and the two $\frac{5}{32}$ in (0.156 in/3·97 mm) steel balls should be inserted into the valve recesses in the casing before the pump body is fitted.
(Note: It is advisable to dismantle the pump for examination before refitting it to the engine. Details of this procedure will be found in Chapter 2).
7 Offer up the pump body, engaging the driving crank (or block on late models) on the pin at the bottom of the distributor shaft. Fit and tighten the securing screws.

31.4 Slide skew gear and crankshaft pinion onto mainshaft

31.5a Fit the tappets into their respective bores

31.5b Place camshaft in position ...

31.5c ... ensuring that timing marks align as shown

31.6 Gasket and ball valves must be positioned before pump is fitted

32 Engine reassembly: fitting the gearbox components, inner casing, kickstart assembly and outer cover

1   If the gearbox mainshaft and layshaft assemblies have been stripped for examination, they should be reassembled, referring to the accompanying line drawing for details. The two clusters should be generously lubricated, and then lowered into the casing together with the selector forks. Note that the gears may not mesh or spin easily at this stage, as only one end of each shaft is supported.

2   The gear change indicator plunger, or Bowden cable should now be attached to the selector quadrant, assuming that one of the two types of indicator is fitted. In the case of the plunger type indicator, feed the operating rod through the hole in the casing, and screw the plunger on to the rod by two or three turns; the plunger body can then be screwed into the casing and tightened.

3   Lower the selector quadrant into the casing, ensuring that the selector fork roller pins engage in their respective grooves. The quadrant will rest on the index spring on the inner face of the casing. If not already in position, offer up the kickstart ratchet shaft, holding the pawl in against spring pressure whilst it is fitted inside the layshaft gear. The gearchange spindle can now be fitted into the casing, retaining the return spring with the special shouldered anchor bolt and shakeproof washer.

4   The inner cover may now be fitted after coating both

mating faces with gasket cement and ensuring that the two locating dowels are in position. Guide the selector quadrant boss into position, then fit its pivot pin and secure it with a split pin. Fit and tighten the inner cover retaining screws. It is a good idea to check that the gearbox operates correctly at this stage. Temporarily refit the gearchange pedal and select each gear in turn. It will be helpful to turn the gearbox sprocket to aid engagement of the gears. On models with a plunger type gear indicator, check that it registers the gear selected.

5   If all appears well, fit the felt sealing washer over the gearchange spindle, and refit the pressed steel cover over the selector quadrant boss.Slide the kickstart return spring over the kickstart spindle, and hook the end of the spring onto the shouldered anchor bolt. Place the return spring plate in position and engage the tang in the return spring. Turn the plate anti-clockwise until the flats on the spindle align with those in the plate. The plate can then be pushed into engagement with the spindle.

6   Offer up the outer casing, having fitted a new gasket.Feed the contact breaker lead through the hole at the rear of the inner casing as the outer casing is positioned. Fit and tighten the securing screws, then refit the gearchange pedal and kickstart crank.

Note: On some machines it will be found that no inspection hole is provided for access to the clutch arm. In this instance, the cover should be left off until the engine unit has been refitted and the clutch cable attached.

32.1a Fit layshaft 4th gear pinion over layshaft end, ...

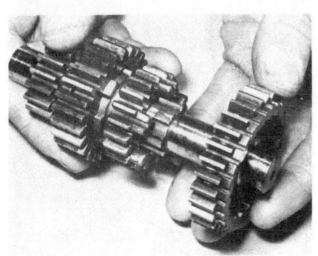

32.1b ... followed by layshaft 1st (low) gear pinion

32.1c Thrust washer must be fitted in 1st gear pinion ...

32.1d ... before kickstart ratchet assembly is inserted

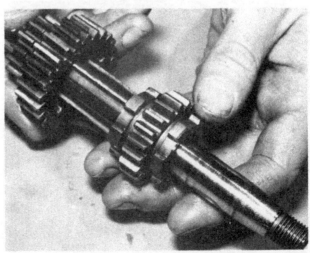

32.1e Slide mainshaft second gear pinion over mainshaft end

32.1f Thrust washer must be positioned inside sleeve gear ...

32.1g ... before gear clusters and forks are lowered into position

32.3a Fit selector quadrant and gearchange spindle assembly

32.3b Gearchange pawls should engage in slots in quadrant

32.3c The return spring is anchored by a shouldered bolt

32.4a Inner cover can now be lowered into position

32.4b Slide the pivot pin through the quadrant boss ...

32.4c ... and retain with split pin

32.5a Felt seal is fitted over gearchange spindle

32.5b Pressed steel cover is retained by two screws

32.5c Kickstart return spring and end plate are fitted as shown

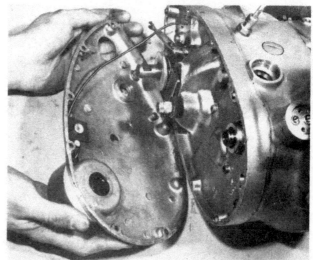

32.6 Guide contact breaker lead through as cover is fitted

33 Engine reassembly: refitting the clutch and primary drive

Note: Details are given in Chapter 3 on fitting a timing pointer on the inside of the chaincase to make timing checks easier. This should be positioned before the primary drive is fitted. See Chapter 3, Section 7 for details.

1   Place the Woodruff key in position in its keyway in the mainshaft end, and then slide the clutch centre into position. Fit a new tab washer, followed by the securing nut. Lock the clutch centre by the method used during its removal, then tighten the nut securely.

2   Check that the clutch outer drum/chainwheel is fitted with its split bronze bearing assembly, ensuring that all the balls are correctly positioned. Fit the primary drive chain around the chainwheel and primary sprocket, and offer the assembly up to their respective shafts.

3   The spacer or distance piece and alternator rotor are fitted next. On most models, the rotor has only one keyway, and should be fitted with the dished face or Lucas motif facing outwards. On machines equipped with energy transfer ignition (no battery fitted), and certain other models, the crankshaft is equipped with two keyways. If the crankshaft is set at TDC, it will be seen that one keyway will be at the 3 o'clock position, whilst the second is at 7 o'clock. On all conventional coil ignition models, the first keyway should be used, the second keyway being specifically for ET (energy transfer) ignition

systems as used on the trials models.

4   Fit the tab washer and nut, holding the crankshaft to prevent its rotation by the same method used during dismantling whilst the nut is tightened. Do not omit to knock over the lock washer.

5   The clutch plates can be inserted next, starting with a plain plate against the friction face in the outer drum/chainwheel unit. If the clutch pushrod has not already been fitted, do so now, before the pressure plate is refitted. Assemble the pressure plate, spring cups and springs, and fit the three screws. On engines up to No 94599, screw the screws inwards until the stud is just flush with the head. On later models, the screw heads should project by about $\frac{3}{8}$ in above the pressure plate.

6   If the engine unit is in position in the frame, reconnect the clutch cable and check that the clutch pressure plate runs evenly. This can be done by operating the clutch lever, holding it in whilst the engine is turned over by engaging top gear and turning the rear wheel. If the pressure plate is seen to wobble as it turns, slacken or tighten the clutch screws as necessary to obtain even rotation. If the unit is still being assembled on the bench, this adjustment should be carried out after the unit has been installed in the frame, as should clutch cable adjustment and pushrod adjustment, which is described in Section 38 of this Chapter.

7   The outer cover should not be refitted at this stage if the ignition timing is to be checked. If the ignition system has not

33.1a Fit Woodruff key to gearbox mainshaft end ...

33.1b ... and lower the clutch centre into position

33.1c Refit and tighten the clutch centre nut

33.2 Clutch outer drum and primary drive are assembled as a unit

33.3a Slide spacer into position on crankshaft end ...

33.3b ... followed by Woodruff key ...

33.3c ... and distance piece

33.3d Alternator rotor can now be fitted and secured

33.5a Fit the clutch plates, starting with a plain plate ...

33.5b ... then a friction plate

33.5c Fit the clutch pressure plate ...

33.5d ... followed by the cups, springs and screws

been disturbed, as in the case of the primary drive or clutch having received attention whilst the engine is in the frame, the cover can be refitted.

34 Engine reassembly: refitting the piston and cylinder barrel

1   Before fitting the piston, it is advisable to pad out the crankcase mouth with clean rag, in order to prevent a displaced circlip from falling in. Extra, unnecessary dismantling work may be called for if the worst happens, and this precaution is not observed.
2   Oil the small end, the piston bosses and the gudgeon pin. Fitting is made easier if the piston is warmed first, especially when a new piston is being fitted after a rebore.
3   As the gudgeon pin is fitted, check that the circlips have engaged with their retaining grooves. A misplaced circlip can work free and cause extensive engine damage, especially if the gudgeon pin is allowed to work out of position and score the cylinder bore.
4   Always use new circlips. It is false economy to use the old

components, even if they appear perfect. There is too much to risk if a circlip breaks or works free whilst the engine is running.
5   The piston should be replaced in the position it was in before removal. A new piston of the flat top type, or one with identical sized valve cutaways may be fitted either way round, unless it is of the split-skirt variety, when incorrect fitment will result in rapid piston and bore wear, and possible engine seizure. Similarly, where the valve cutaways are of differing sizes, the piston must be placed so that the larger cutaway corresponds with the inlet valve, which has a larger head than the exhaust valve.
6   Fit a new cylinder base gasket, oil the cylinder bore with clean engine oil and arrange the piston rings so that their end gaps are approximately 120° apart on the piston. Rest the piston on two strips of wood placed across the crankcase mouth, then lower the cylinder barrel into position.
7   The piston rings may be compressed, one at a time, as the lowering of the barrel is accomplished. There is a tapered lead-in at the base of the cylinder bore that will facilitate fitting. When the oil control ring has entered the bore fully, remove the supporting wood strips followed by the packing rag. Push the cylinder barrel downwards until it seats on the crankcase.

34.3 Pad crankcase mouth with rag, then fit piston, using new circlips

34.7 Oil piston and bore, then slide barrel into position

35 Engine reassembly: setting the ignition timing

1   It is recommended that the ignition timing be checked at this stage as it is advantageous to be able to see the piston movement, which would otherwise be obscured by the cylinder head. The adjustment sequence will be found in Chapter 3, together with details of constructing a pointer system to make subsequent checks easier.

36 Engine reassembly: refitting the cylinder head

1   Fit new sealing rings to the pushrod tube, noting that the off-white silicone rubber seal fits at the top of the tube, the black seal fitting into the recess at the bottom. Note that three types of pushrod tube have been used; the earliest type had a small cutout at the bottom of the tube which engages in a corresponding raised pip on the crankcase. A second type had alignment marks on the cylinder head and at the top of the tube. Both these types had a guide plate near the top of the tube which locates the pushrods, hence the need for some sort of alignment. Later machines dispensed with the guide plate, and

consequently there is no need to align the tube during assembly.

2   When fitting the pushrod tube and pushrods, note that the inner tappet operates the exhaust valve, and the outer tappet the inlet valve. This is normally marked on the tappet boss. On late models with no guide plate in the pushrod tube, the pushrods may be held in place with a small dab of grease. Ensure that the pushrod tube seats squarely in its recess.

3   The copper cylinder head gasket may be re-used if it is undamaged, having first annealed it by heating it to a dull cherry red colour, and then quenching it in cold water. It is good practice to lightly grease the sealing faces of the gasket before placing it on the mating face of the cylinder barrel. Check that the engine is set at TDC compression, at which point the piston will be at the top of its stroke, and both pushrods will be level.

4   Lower the cylinder head into position, ensuring that the pushrods engage on the rocker ball ends. A small screwdriver may be used to guide the pushrods into position. Fit the four cylinder head sleeve nuts, and tighten them down evenly, in a diagonal sequence. No torque settings are specified, so some discretion on the part of the owner must be exercised. Provided that the nuts and studs are in good order, no damage will be done by using firm hand pressure on normal pattern spanners.

36.3 Place cylinder head gasket over studs – note locating dowels

36.4a Lower cylinder head over holding down studs ...

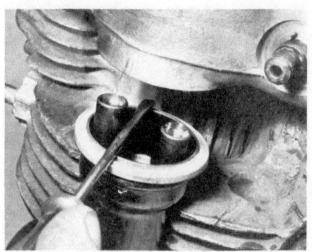

36.4b ... using a small screwdriver to guide pushrods into place

37 Fitting the engine unit into the frame

1   Refitting the engine unit is basically a straightforward reversal of the removal sequence. Check that there are no trailing leads or cables which might snag on the engine as it is positioned. It is a good idea to enlist the aid of an assistant to steady the frame and to help position the unit. Tilt the crankcase down at the rear, guiding the front lug past the frame lugs. When the engine is in position, slide the three mounting bolts into place, then fit and tighten the securing nuts.

2   Feed the clutch cable into position through the rear of the timing case, then engage the nipple in the forked end of the clutch actuating arm. If the cover is of the type which does not have a clutch arm inspection plug, it should now be refitted, as should the gearchange lever and kickstart pedal.

3   Refit the carburettor and reconnect the throttle cable. Where an Amal Type 32 instrument is employed, ensure that the float chamber is kept level, i.e. that the instrument is set vertically, and that the clamp is not overtightened. Use a new paper gasket on the flange to maintain an airtight joint. On no account should silicone-rubber 'plastic gasket' compounds be used on carburettor joints, as this compound is not resistant to direct contact with petrol.

4   Reconnect the oil pipes at the junction block on the

underside of the crankcase, using a new gasket. Note that it may prove necessary to remove the engine lower mounting bolt to provide clearance. Reconnect the pipes at the tank, if these have been disturbed. The return feed stub on the tank can be identified by the take-off pipe to the rocker feed.

5   Fit an annealed $\frac{3}{8}$ in copper washer to the end of each rocker spindle, followed by the two banjo unions, and two annealed $\frac{5}{16}$ in copper washers. Fit the domed nut to each spindle, then tighten each nut sufficiently to form an oil tight joint. Use pliers or a self-grip wrench to hold the unions whilst tightening the nuts. If this is not done, the feed pipe may bend and fracture due to the union being twisted.

6   Fit the exhaust pipe and clamp over the exhaust port stub, and secure the silencer to the frame with the right-hand pillion footrest. Position the exhaust port clamp and tighten the pinch bolt to retain the pipe.

7   Fit the final drive chain and secure the joining link, ensuring that the closed end of the spring link faces the direction of travel. The chainguard can now be refitted, if this was removed, as can the footrests and rear brake pedal. Check the adjustment of the rear brake after setting the rear wheel to give about 1 in up-and-down play in the drive chain.

8   Remove the gearbox filler plug, and add SAE 30 engine oil to the level of the level tube in the gearbox drain plug. Refit the small concentric level plug. The gearbox contains approximately 200 cc ($\frac{1}{3}$ pint) of oil. Fill the engine oil tank to the level line with SAE 20 or 30 engine oil. If the level mark is missing or has been obliterated, fill the tank to within about $1\frac{1}{2}$ in of the filler neck.

9   Reconnect the alternator output leads and the contact breaker leads at their relevant bullet connectors. The cables are colour-coded to aid identification. On machines equipped with a remote air filter, refit the rubber connecting pipe between the filter assembly and carburettor mouth.

38 Final adjustments

Clutch adjustment

1   Remove the primary chain cover inspection screw and slacken the clutch adjuster locknut. Turn the adjuster inwards until the clutch actuating arm is felt to have touched the inside of the cover. On late models, this can be observed via the inspection plug. Back off the adjuster by $\frac{1}{2}$ to 1 turns to give a reasonable amount of free play in the clutch actuating mechanism, then tighten the locknut whilst holding this setting. Note that failure to ensure adequate free play may cause clutch pushrod wear and clutch slip, so do not overlook this stage.

2   Adjust the clutch cable adjuster to give about $\frac{1}{16}$ in free play in the cable. Withdraw the clutch lever and turn the engine over, whilst watching the clutch pressure plate. If any sign of wobble is present, it will be necessary to slacken or tighten the clutch screws to obtain even running. If left unattended, wobble in the clutch pressure plate will result in indifferent clutch operation, and may cause judder.

Valve rocker clearance

3   Set the engine at TDC compression. Slacken the adjuster locknuts and back off the adjusters. On most engines, the correct valve rocker clearance is 0·010 in (0.25 mm) for both the inlet and exhaust valves, but on some competition engines, and on any engine fitted with a sports camshaft, this setting becomes 0·002 in (0.5 mm) inlet and 0·004 in (0.10 mm) exhaust. Insert the relevant feeler gauge, and turn the adjuster by hand until it just nips the gauge, giving a light sliding fit. Hold this setting and tighten the locknut. This procedure should be repeated on the remaining valve, after which the engine should be turned over once or twice and the settings re-checked.

4   It is worth knowing that each quarter turn of the adjuster screw corresponds to 0·010 in (0.25 mm) clearance, and this method can be used to set the clearance in an emergency. However, it is advisable to use a feeler gauge whenever possible.

5   Liberally lubricate the rockers and pushrods before fitting

the covers, using new gaskets and new or annealed 'O' rings. Fit and tighten each of the rocker cover retaining nuts, holding the covers in position.

General checks and adjustments

6   Check round the unit for any loose fittings, leads or cables. Fit and reconnect the battery (+ earth) and check that the lights and horn operate correctly. Check that the various controls are correctly adjusted and that they function properly.

7   Refit the rear fairing halves, where applicable, and fit the petrol tank and seat. Connect the petrol feed pipe and check that no leaks are present. Connect the high tension lead to the sparking plug.

39 Starting and running the rebuilt engine

1   When the engine starts, remove the oil tank filler cap and check that oil is returning. There may be a time lag before oil commences to emerge from the return pipe because pressure has to build up in the rebuilt engine before circulation is complete, but do not permit the engine to run at low speed for more than a couple of minutes before stopping it and checking the system.

2   The exhaust will smoke excessively during the initial start, due to the presence of oil used throughout the reassembly process. This should gradually disperse as the engine settles down.

3   The return to the oil tank will eventually contain air bubbles because the scavenge pump will have cleared the excess oil content of the crankcase. The scavenge pump has a greater capacity than the feed pump, hence the presence of air when there is little oil to pick up.

4   Check the engine for leakages at gaskets and pipe unions etc. It is unlikely any will be evident if the engine has been reassembled correctly, with new gaskets and clean jointing faces. Before taking the machine on the road check that both brakes work effectively and that all controls operate freely.

5   If the engine has been rebored, or if a number of new parts have been fitted, a certain amount of running-in will be required. Particular care should be taken during the first 100 miles or so, when the engine is most likely to tighten up, if it is overstressed. Commence by making maximum use of the gearbox, so that only a light loading is applied to the engine. Speeds can be worked up gradually until full performance is obtained with increasing mileage.

6   Do not tamper with the silencer or fit another design unless it is designed specifically for a Tiger Cub. A noisier exhaust does not necessarily mean improved performance; in a great many instances unwarranted modifications or the fitting of an unsuitable design of silencer will have an adverse effect on both performance and petrol consumption.

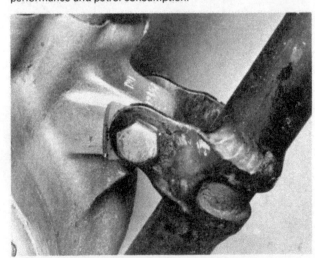

37.1a Fit engine front mounting bolt ...

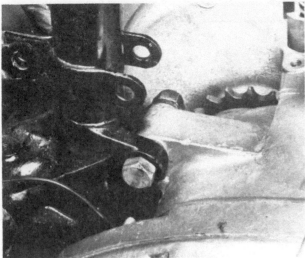

37.1b ... and engine rear mounting bolt

37.1c Fit oil feed manifold before lower mounting bolt is tightened

37.3 Fit carburettor using new gaskets

37.5 Connect oil feed pipes to rocker shaft ends

37.8 Top up the gearbox to level plug height

38.3 Check and adjust valve clearances before refitting covers

40 Fault diagnosis: engine

Symptom	Cause	Remedy
Engine will not start	Defective sparking plug	Remove plug and lay it on cylinder head. Check whether spark occurs when engine is kicked over.
	Dirty or closed contact breaker points	Check condition of points and whether gap is correct.
	Ignition coil defective	Remove HT lead from plug and jam between cylinder head fins, take off C/B cover, switch on ignition and using an insulated screwdriver flick the C/B points open – a fat consistent spark should result.
	Fuel starvation	Check fuel supply. Check to see that the fuel tap is turned on. Check the fuel lines for obstruction.
	Fuel flooding	Remove and dry sparking plug.
	Low compression	If the engine can be turned over on the kickstart with less than normal effort, perform a compression test and determine the cause of low compression. (To check compression buy or borrow a test gauge of the type that is screwed or held in the sparking plug hole while while the engine is kicked over).
	Pushrods incorrectly fitted	Dismantle and reassemble as described in Section 36
Engine runs unevenly	Ignition system fault	Check system as though engine will not start.
	Blowing cylinder head gasket, or bad sealing joint	Leak should be evident from oil leakage where gas escapes.
	Incorrect ignition timing	Check timing and reset if necessary.
	Incorrect fuel mixture	Adjust carburettor. Remove and clean carburettor jets. Check float level. Check for intake air leaks. Make sure that the carburettor mounting bolts are tight.
Lack of power	Incorrect ignition timing	See 'Ignition timing' in Chapter 3.
	Fault in fuel system	Check system and filler cap vent.
	Blowing head gasket	See above.
High oil consumption	Cylinder barrel in need of rebore and o/s piston	Fit new rings and piston after rebore.
	Oil leaks or air leaks from damaged gaskets or oil seals	Trace source of leak and replace damaged gaskets or seals.
	Oil not returning to tank	Remove oil tank filler cap and check whether oil is flowing from the return pipe whilst the engine is running. If trouble persists check oil pump.
Excessive mechanical noise	Worn cylinder barrel (piston slap)	Rebore and fit o/s piston.
	Worn small end bearing (rattle)	Renew bearing and gudgeon pin.
	Worn big-end bearing (knock)	Fit new big-end bearing.
	Worn main bearings (rumble)	Fit new journal bearings.
Engine overheats and fades	Pre-ignition and/or weak mixture	Check carburettor settings. Check also whether plug grade correct.
	Lubrication failure	Check operation of pump as above, also check for possible oilway/oil pipe blockage.

41 Fault diagnosis: clutch

Symptom	Cause	Remedy
Engine speed increases but not road speed	Clutch slip; incorrect adjustment or worn linings	Adjust, or renew clutch plates.
Machine creeps forward when in gear; difficulty in finding neutral	Clutch drag; incorrect adjustment or damaged clutch plates	Readjust, or fit new clutch plates.
Machine jerks on take-off or when changing gear	Clutch centre loose on gearbox mainshaft	Check for wear and retighten retaining nut.
Clutch noisy when withdrawn	Badly worn clutch centre bearing	Renew bearing.
Clutch neither frees nor engages smoothly	Burrs on edges of clutch plates and slots in clutch drums	Dress damaged parts with file if damage not too great.
Clutch action heavy	Overtight tension springs or wrong angle of operating arm Dry operating cable, or bends too tight	Slacken tension nuts or readjust operating arm. Lubricate cable and reroute as necessary.
Clutch 'bites' at extreme end of lever movement	Worn linings	Renew clutch plates (inserted).
Constant loss of clutch adjustment	Worn pushrod, due to failure to maintain minimum clearance.	Renew pushrod and readjust.

42 Fault diagnosis: gearbox

Symptom	Cause	Remedy
Kickstart does not return when engine is turned over or started	Broken kickstart return spring	Renew.
Kickstart slips and will not turn engine over	Worn ratchet assembly	Renew.
Kickstart jams	Worn ratchet assembly and/or kickstart quadrant	Renew.
Difficulty in engaging gears	Selector pawls not engaging due to wear Weak gearchange mechanism springs Selector forks bent or badly worn	Renew pawls. Renew springs. Renew.
Machine jumps out of gear	Worn dogs on gear pinions	Renew defective pinions.
Gearchange lever does not return to original position	Broken return spring Weak gearchange mechanism springs	Renew spring. Renew springs.

Chapter 2 Fuel system and lubrication

Contents

General description 1
Petrol tank: removal, examination and replacement 2
Petrol tap: removal, examination and replacement 3
Petrol feed pipe: examination 4
Carburettor: removal 5
Carburettor: dismantling, examination and reassembly . . . 6
Carburettor: checking the settings 7
Air filter element: location, examination and cleaning 8
Exhaust system 9
Checking the lubrication system 10
Oil pump: general description 11
Oil pump: removal, examination and maintenance 12
Rocker oil feed: examination and renovation 13
Fault diagnosis 14

Specifications

Petrol tank capacity

T.15 Terrier
T.20 up to engine No. 17388
T.20C and T.20T
T.20S and T.20SS . . . . $2\frac{5}{8}$ gall (11·9 litres)
T.20 from 17388 to 35846 . . . . $3\frac{1}{8}$ gall (14·2 litres)
T.20 from 35846 onwards
T.20SS and T.20SH . . . . 3 gall (13.5 litres)

Oil tank capacity

T.15 Terrier and T.20 to 17388 . . . . $2\frac{1}{4}$ imp pint (1·4 litres)
Other models . . . . $2\frac{3}{4}$ imp pint (1·56 litres)

Primary chaincase capacity

T.20 from 56360
T.20T and T.20S
T.20SS and T.20SH . . . . $\frac{1}{3}$ imp pint (200 cc)
Other models . . . . $\frac{1}{2}$ imp pint (300 cc)

Gearbox capacity . . . . $\frac{1}{3}$ imp pint (200 cc)

Carburettor

Model/Year	Terrier/1953	Terrier 1954-6	T.20 1954-7	T.20 1957	T.20/J 1957-8
Type (Amal)	332/1	332/2	332/3	332/4	332/5
Main jet	120	90	100	100	100
Needle jet	.086	.086	.086	.086	.086
Needle position	2	3	3	3	3
Throttle valve	4	4	4	4	4
Pilot jet	-	20	20	20	20
Choke size	$\frac{3}{4}$ in	$\frac{11}{16}$ in	$\frac{3}{4}$ in	$\frac{3}{4}$ in	$\frac{3}{4}$ in

Model/year	Terrier/1958	Export T.20 1959	T.20S 1959-60	T.20S,SL,SS,SH 1961-6	T.20 and T 1960-1
Type (Amal)	332/6	332/7	376/217	376/272	375/44
Main jet	90	140	140	140	100
Needle jet	.086	.086	.106	.106	.105
Needle position	3	3	3	3	3
Throttle valve	4	4	3	$2\frac{1}{2}$	$3\frac{1}{2}$
Pilot jet	20	15	20	20	25
Choke size	$\frac{11}{16}$	$\frac{13}{16}$	$\frac{15}{16}$	$\frac{15}{16}$	$\frac{25}{32}$

	T.20 and trials	T.20 SM	T.20C and SC
Model/year	1963-6	1965-8	1967-8
Type (Amal)	376/44	376/314	375/61
Main jet	100	140	90
Needle jet	.105	.105	-
Needle position	3	3	3
Throttle valve	$3\frac{1}{2}$	$3\frac{1}{2}$	$3\frac{1}{2}$
Pilot jet	25	15	25
Choke size	$\frac{25}{32}$	$\frac{15}{16}$	$\frac{25}{32}$

Model	T.20 and C	T.20 and T
Type (Zenith)	17 MX	18 MX
Main jet	78	84
Slow running jet	50	45
Starter slide	200/65	200/65
Choke size	17 mm	18 mm

1 General description

1 The fuel system comprises a petrol tank from which petrol is gravity fed to the float chamber of the carburettor via a petrol tap and gauze filter element.

2 A number of types of carburettor were fitted to the Tiger Cub models, varying according to the year of manufacture, model and application. Early models were fitted with Amal instruments, and as a general rule, this was the case with subsequent models. For a short period, Zenith 17MX and 18MX instruments were employed. T.20 and T.20C models between engine numbers 35846 and 56360 used the 17MX, the later T.20 models and the T.20.T using the larger 18MX. In practice, these carburettors were often discarded in favour of the contemporary Amal instruments, these being available as replacements. This was mainly due to the problems of repairing or overhauling a worn Zenith unit, as many of its component parts were fixed permanently in position. Do not be surprised if the instrument listed in the Specifications does not match that fitted to your machine. The various instruments were often changed in an attempt to obtain more power.

3 Lubrication on all models is effected on the dry sump principle, in which the oil is gravity-fed from a side mounted oil tank to a twin plunger type oil pump mounted on the right-hand side of the crankcase. The smaller of the two pump cylinders forces oil through the hollow mainshaft end to the big-end bearing. Lubricant escaping from the bearing is splashed on the cylinder walls and small end eye, and then drains down into the sump.

4 The larger pump plunger is used to pump the oil from the crankcase via a gauze filter, to the oil tank. A junction at the oil tank union allows a proportion of the returning oil to be fed at reduced pressure to the rocker shafts. This oil then runs down the pushrod tunnel, lubricating the cam followers and the camshaft and pinions, eventually draining back into the crankcase.

2 Petrol tank: removal, examination and replacement

1 Turn the petrol tap to the 'off' position, and disconnect the petrol pipe, either at the tap, or at the carburettor union. Slacken and remove the two petrol tank mounting bolts, and lift the tank away.

2 Filler cap seals may be made up from synthetic rubber sheet cut to size. On no account should natural rubber sheet be used, as this is attacked by petrol. The tank is refitted by reversing the removal sequence.

3 Petrol tap: removal, examination and replacement

1 The petrol tap is a simple 'pull for on' plunger type, having no reserve setting. The tap body screws into a threaded boss in the underside of the tank, and can be removed for examination and access to the gauze petrol filter.

2 Drain the petrol, or remove the tank and place it on its side so that the tap orifice is above the level of the petrol. The tap may now be unscrewed from the boss. Remove any accumulations of sediment from the filter gauze, using clean petrol.

3 The tap plunger is located by a small screw in the side of the tap body. When this has been removed, the tap plunger can be pulled out of the body for examination. The seal is formed by a cylindrical cork section which covers the feed hole when not in use. If the cork becomes scored or badly worn, leakage will occur, necessitating renewal. Unfortunately, new plunger assemblies are becoming increasingly difficult to locate. A possible alternative is a sound secondhand item, or a complete lever-type tap of the same fitting.

4 If the machine has been left unused for any length of time, the cork may have shrunk, leading to leakage when used again. The remedy in this case is to immerse the tap plunger in boiling

3.3 A worn cork seal can allow leakage to occur

water for a few minutes, to swell the cork sufficiently to seal against the tap body. Alternatively, the cork can be immersed in castor oil, which will produce a similar result.

4 Petrol feed pipe: examination

1 The petrol feed pipe is connected to a banjo union at the carburettor end, and to a stub on the petrol tap. The pipe is of transparent plastic tubing. It will be noticed that the tubing becomes yellow with age, and eventually becomes brittle as the plasticiser is leached out by the petrol. A pipe in this condition should be renewed as a precaution against cracking.

2 Synthetic rubber tubing may also be used, which is more resistant to ageing. It can, however, give some trouble due to particles of rubber breaking away from the inner walls of the pipe, and causing obstructions in the filter or jets. On no account should natural rubber tubing be used, even in an emergency, as this is attacked by petrol and will quickly block the carburettor.

5 Carburettor: removal

1 The carburettor is mounted either directly by a flange fixing, or by way of an adaptor, to which it is secured by a pinch bolt. Before attempting to remove the instrument, turn the petrol tap to the off position, and disconnect the petrol feed pipe. Detach the carburettor to air filter hose, where fitted.

2 Remove the mixing chamber top by unscrewing the knurled retaining ring, or by releasing the retaining screw(s), depending upon the type of carburettor fitted. The throttle valve assembly can now be withdrawn, and the throttle cable disengaged if it is thought necessary. If the assembly does not require attention, it may be left attached to the cable.

3 Slacken the pinch bolt, or remove the flange mounting nuts as applicable, and remove the carburettor assembly. Note that it is not necessary to remove the manifold adapter.

6 Carburettor: dismantling, examination and reassembly

Zenith MX17 and MX18

1 Remove the float bowl retaining screws, and lift the bowl away, taking care not to damage the gasket, which may be reused if unbroken and not compressed. Lift out the float, noting that the pivot pin may drop out and is easily lost.

2 The main jet and emulsion tube assembly can be unscrewed from the underside of the body, as can the smaller

slow running jet. If still in position, release the petrol pipe banjo union so that the gauze filter element may be cleaned. The pilot mixture strength is controlled by the slow running jet, and no adjustment screw is provided. There is no need to disturb the throttle stop screw.

3   A small gauze air filter is normally fitted to the intake side of the instrument and is retained by two screws. It can be removed and cleaned by washing in petrol to remove accumulated dust. Blow dry with compressed air, and oil the gauze before refitting the filter to the carburettor.

4   A brass plunger is incorporated in the carburettor body, which when depressed provides an enrichened mixture to aid cold starting. It is returned by the throttle slide as this is opened. This device requires no maintenance. The float needle will be found to be fixed in position, and therefore removal for cleaning or renewal is not feasible. If flooding has been a problem, it can only be hoped that this is due to dirt on the needle, which can be flushed off, rather than advanced wear, which normally means renewal of the carburettor body.

5   When refitting the instrument to the inlet adaptor, ensure that the float bowl is kept upright to prevent the float level becoming upset. Check also that the 'O'ring which seals the joint is in good order, or air leaks and poor slow running will result.

Amal Type 32

6   These carburettors are very similar in design to the Zenith MX series, and most of the comments in the preceding sub-section can be applied. Note that the float bowl has a third retaining screw. The pilot jet and needle jet (or emulsion tube) are screwed into the flange of the float bowl, the main jet being fitted beneath a brass plug in the bottom of the float bowl. Note that the float needle will drop free on this model, and may be renewed, if worn.

7   Mixture strength, for slow running purposes, can be adjusted by way of the vertical screw on the right-hand side of the body, the horizontal screw forming the throttle stop adjustment. The starting, or choke slide normally requires no attention, but may be dismantled if required, referring to the accompanying photographs for details. When refitting the carburettor, ensure that the insulating sleeve is fitted, and is in good condition.

Amal Standard

8   This type of instrument differs from the Zenith and Amal 32 types, and is of traditional Amal design., The float chamber top is secured by two hexagon-headed screws. The brass float is fitted with a long needle through its centre, the conical end of which seats in a valve in the base of the chamber. The float chamber incorporates a lug which forms a banjo union, by which the float chamber is joined to the main body, and which channels petrol between the two.

9   Access to the main jet and needle jet is gained by removing the large securing bolt and the float chamber, the jets being screwed into the base of the body. The pilot jet is screwed into the underside of the choke, and is covered by a chromium plated domed nut.

10   The throttle stop adjustment is controlled by a screw and locknut which passes at an angle into the mixing chamber. Pilot mixture adjustment is by way of the spring loaded screw which passes horizontally into the carburettor body.

11   Care must be taken to renew any suspect sealing washers during reassembly, otherwise leakage will result, particularly at the union between the main body and float chamber. Check that the float chamber is positioned correctly, so that it is vertical when the carburettor is refitted.

Amal Monobloc

12   The Monobloc carburettor is so named, as it was the first instrument manufactured by Amal Limited that incorporated the float chamber as an integral part of the main carburettor body. To gain access to the float and float needle, it is necessary to remove the circular cover that forms the side of the float chamber housing. It is retained by three screws and has a gasket on the inside joint. The float hinges on a pin projecting from the float chamber wall. It is preceded by a small brass spacer. When the float is withdrawn, the nylon float needle will be displaced from its seating and fall clear.

13   The jet block is retained by a screw close to the pilot jet adjusting screw, and by a large hexagon nut at the base of the mixing chamber. Before the latter can be unscrewed, the hexagon bolt below it should be removed first and the main jet and needle jet unscrewed from the jet block. The main jet threads into the lower end of the needle jet, which itself screws into the mixing chamber base. When the large hexagon nut has been removed and the small screw in the side of the mixing chamber, the jet block can be drifted out of position in an upwards direction, using great care because it is made of brass.

14   The throttle slide, needle and air slide assembly will remain attached to the mixing chamber top. If it is necessary to dismantle this assembly, retract the air slide by operating the handlebar control and disengage it from the slot in the top of the throttle slide. Remove the throttle cable from the throttle slide by withdrawing the split pin that retains the lower cable nipple in its seating, then raise the slide upwards against the pressure of the return spring so that the cable can be disengaged completely. Take off the return spring and place it in a safe place for reassembly.

15   The needle is retained by a spring clip. Before withdrawing the clip to release the needle, note the needle position. It has five notches to give variation of mixture strength and must be replaced in the same position.

16   It is unlikely that the air slide assembly will need to be dismantled. If, however, such action is necessary, displace the lower cable nipple from its seating in the base of the slide, so that it protrudes through the slot in the slide body. The slide can be pulled off the cable, followed by the shouldered guide on which it seats, when raised. Do not misplace the return spring.

17   If it is necessary to dismantle the float chamber tickler, access is gained by unscrewing the hexagon nut that surrounds the tickler plunger. The Monobloc carburettor has a detachable pilot jet. It is housed in the underside of the mixing chamber body, close to the flange joint and is blanked off by a hexagon headed plug. If this plug is removed, the threaded pilot jet within the carburettor body can be unscrewed. It has a slotted end, to facilitate removal with a screwdriver.

Examination

18   Check the float to see whether it has become porous and allows petrol to enter and upset its balance. Irrespective of whether the float is made of copper or plastic, it should be replaced if a leak is evident. It is not practicable to effect a satisfactory repair.

19   Check the float needle and float needle seating to see whether the float needle is bent or whether a ridge has worn around either the needle or its seating as the result of general wear. All defective parts should be replaced. The needle seat will unscrew from the float chamber body, to permit replacement when necessary.

20   The throttle slide, needle and air slide assembly still attached to the carburettor top should be examined. Signs of wear on the throttle slide will be self-evident, if the amount of wear is particularly high it may be responsible for a pronounced clicking noise when the engine is running slowly, as the slide moves backwards and forwards within the mixing chamber.

21   The needle should be straight and the needle retaining clip a good fit. Check the needle for straightness by rolling it on a sheet of plate glass. If it is bent, it must be replaced. Reject any needle clip that has lost its tension.

22   The air slide assembly seldom requires attention. Trouble can occur if the compression spring loses its tension, since this will cause the air slide to stick, making cold starting more difficult.

23   Check the main jet, needle jet and pilot jet (if fitted). Wear will occur in the needle jet only; the other jets are liable to blockages if dirty or contaminated petrol is used. NEVER use

wire to clear a blocked jet otherwise there is danger of the hole being enlarged. Use either a foot pump or a compressed air line to clear the blockage.

24 The 'Monobloc' carburettor fitted to some models has a synthetic rubber 'O'ring in the centre of the mounting flange. This seal must be in good condition to prevent air leaks.

25 On all carburettors there is a pronounced tendency for the mounting flange to 'bow', if the retaining nuts are over-tightened. The resultant air leak, which will have a marked effect on carburation, will be difficult to trace as a result. The condition of the flange can be checked by holding a straight edge across the face. If the flange is bowed, it should be rubbed down with a sheet of fine emery cloth wrapped around a piece of flat glass, using a rotary motion, until the bow is removed. Make sure the carburettor body is washed very thoroughly after this operation, to ensure that small particles of abrasive do not lodge in any of the small internal air passages.

All carburettors

26 To reassemble the carburettor, reverse the procedure used for dismantling. Note that undue force should never be used during dismantling or reassembly because the castings are usually in a zinc-based alloy which will fracture very easily if overstressed. Do not omit the float pivot spacer from the Monobloc carburettor, or persistent flooding will occur.

5.1 The carburettor may be flange mounted, or clamped as shown

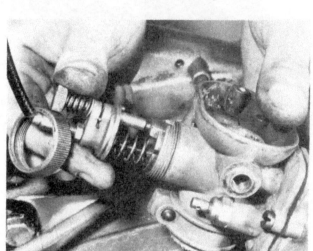

5.2 Throttle valve assembly should be removed complete

6.6a Float bowl is retained by one screw here ...

6.6b ... and by two screws on this side

6.6c Pilot jet and emulsion tube screw into float bowl

6.6d Main jet is fitted beneath brass plug in float bowl

6.6e Float and pivot pin can be lifted away

6.6f Float needle will drop free – take care not to lose it

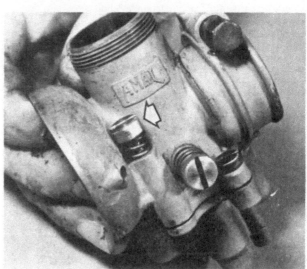

6.7a This screw adjusts the pilot mixture strength

6.7b Horizontal screw controls throttle stop position

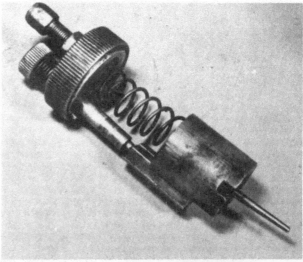

6.7c Mixing chamber components are arranged thus

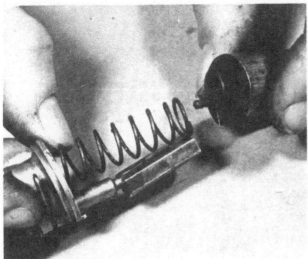

6.7d Throttle valve and needle can be released after displacing throttle cable

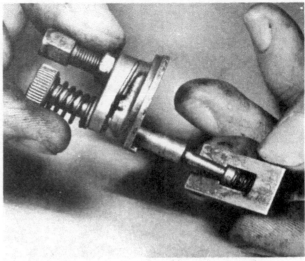

6.7e Choke slide can be disengaged from plunger rod ...

6.7f ... and spring clip displaced, freeing plunger

6.7g Choke assembly components

6.7h Check that all joints are airtight during reassembly

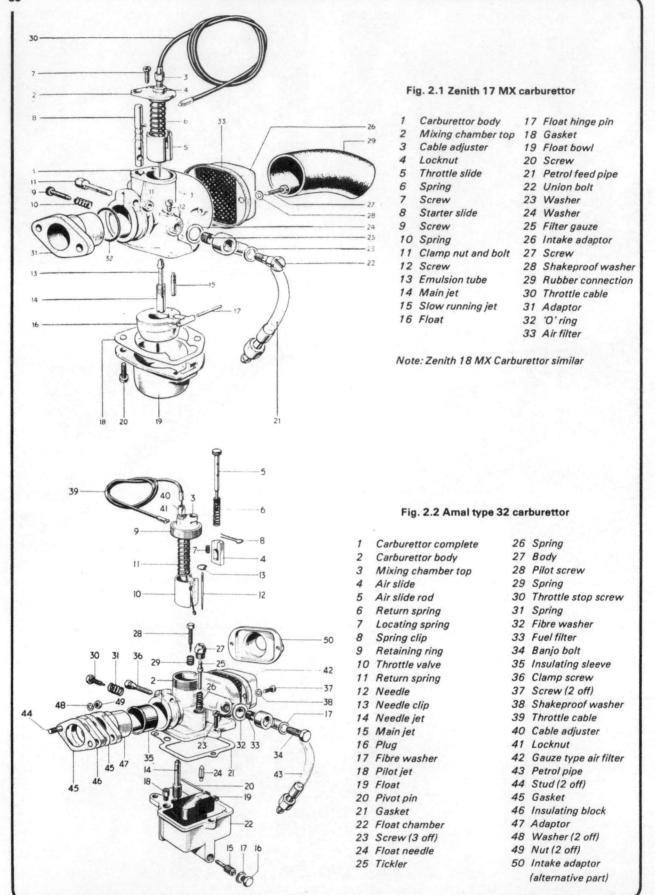

Fig. 2.1 Zenith 17 MX carburettor

1	Carburettor body	17	Float hinge pin
2	Mixing chamber top	18	Gasket
3	Cable adjuster	19	Float bowl
4	Locknut	20	Screw
5	Throttle slide	21	Petrol feed pipe
6	Spring	22	Union bolt
7	Screw	23	Washer
8	Starter slide	24	Washer
9	Screw	25	Filter gauze
10	Spring	26	Intake adaptor
11	Clamp nut and bolt	27	Screw
12	Screw	28	Shakeproof washer
13	Emulsion tube	29	Rubber connection
14	Main jet	30	Throttle cable
15	Slow running jet	31	Adaptor
16	Float	32	'O' ring
		33	Air filter

Note: Zenith 18 MX Carburettor similar

Fig. 2.2 Amal type 32 carburettor

1	Carburettor complete	26	Spring
2	Carburettor body	27	Body
3	Mixing chamber top	28	Pilot screw
4	Air slide	29	Spring
5	Air slide rod	30	Throttle stop screw
6	Return spring	31	Spring
7	Locating spring	32	Fibre washer
8	Spring clip	33	Fuel filter
9	Retaining ring	34	Banjo bolt
10	Throttle valve	35	Insulating sleeve
11	Return spring	36	Clamp screw
12	Needle	37	Screw (2 off)
13	Needle clip	38	Shakeproof washer
14	Needle jet	39	Throttle cable
15	Main jet	40	Cable adjuster
16	Plug	41	Locknut
17	Fibre washer	42	Gauze type air filter
18	Pilot jet	43	Petrol pipe
19	Float	44	Stud (2 off)
20	Pivot pin	45	Gasket
21	Gasket	46	Insulating block
22	Float chamber	47	Adaptor
23	Screw (3 off)	48	Washer (2 off)
24	Float needle	49	Nut (2 off)
25	Tickler	50	Intake adaptor
			(alternative part)

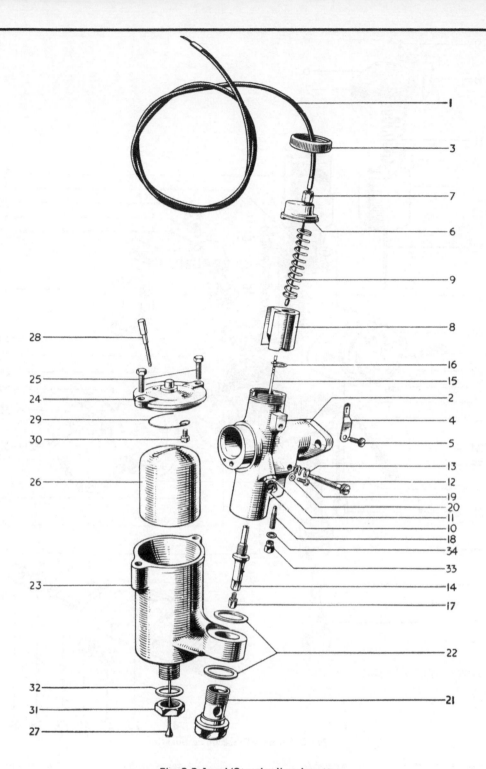

Fig. 2.3 Amal 'Standard' carburettor

1	Throttle cable	9	Spring	17	Main jet	26	Float
2	Mixing chamber body	10	Adjusting screw	18	Pilot jet	27	Float needle
3	Cap	11	Nut	19	Screw	28	Tickler
4	Cap spring	12	Air adjusting screw	20	Washer	29	Spring
5	Screw	13	Spring	21	Holding bolt	30	Fixing screw
6	Mixing chamber top	14	Needle jet	22	Washer – 2 off	31	Locknut
7	Throttle cable adjuster	15	Jet needle	23	Float chamber body	32	Washer
8	Throttle valve	16	Clip	24	Float chamber cover	33	Cover nut
				25	Screw – 2 off	34	Washer

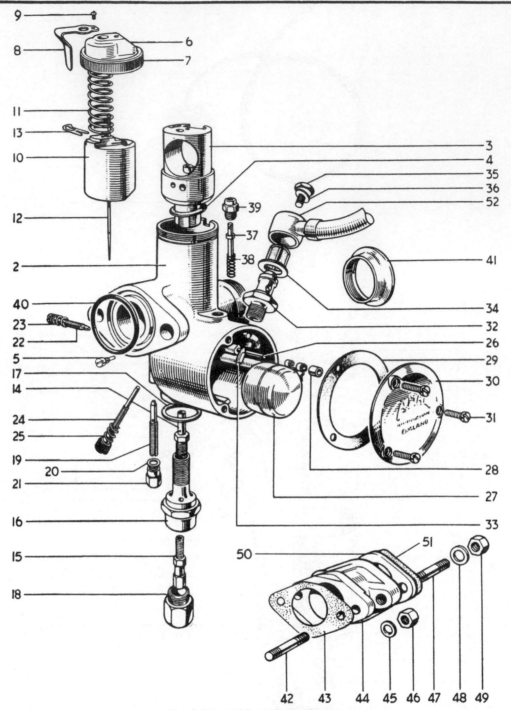

Fig. 2.4 Amal Monobloc carburettor

1	Carburettor complete	14	Needle jet	27	Float assembly	40	O ring
2	Carburettor body	15	Main jet	28	Spacer	41	Bellmouth
3	Jet block	16	Jet holder	29	Gasket	42	Stud (2 off)
4	Sealing washer	17	Fibre washer	30	Float chamber cover	43	Gasket
5	Guide screw	18	Cap nut	31	Screw (3 off)	44	Adaptor
6	Mixing chamber top	19	Pilot jet	32	Union adaptor	45	Washer (2 off)
7	Retaining ring	20	Fibre washer	33	Float needle	46	Nut (2 off)
8	Locking spring	21	Cap nut	34	Fuel filter	47	Stud (2 off)
9	Screw	22	Pilot screw	35	Banjo bolt	48	Washer (2 off)
10	Throttle valve	23	Spring	36	Fibre washer	49	Nut (2 off)
11	Return spring	24	Throttle stop screw	37	Tickler	50	Insulating block
12	Needle	25	Spring	38	Spring	51	Gasket
13	Needle clip	26	Pivot pin	39	Body	52	Feed pipe union

7 Carburettor: checking the settings

1 The various sizes of jets and that of the throttle valve and needle are predetermined by the manufacturer and should not require modification. Refer to the Specifications list if there is any doubt about the values fitted. It should be noted that carburettor settings on all competition machinery should be made in reference to particular combinations of exhaust pipe length and silencer type.

2 Slow running is controlled by a combination of the throttle stop and air regulating (pilot jet) screw settings, on all but the Zenith MX instruments. These have a fixed slow running jet, which cannot be adjusted. On all other carburettors, the pilot mixture should be set as follows. Commence by screwing the throttle stop screw inwards so that the engine runs at a fast tickover speed. Adjust the air screw setting until the tickover is even, without misfiring or hunting. Readjust the throttle stop until the desired tickover speed is obtained, then re-check the air regulating screw so that the tickover is as even as possible. Remember that turning the pilot adjuster screw out weakens the mixture. Always make these adjustments with the engine at normal running temperature and remember that the more sporting models having a high degree of valve overlap are unlikely to run very evenly at low speeds, no matter how carefully the adjustments are made. In this latter category, it is often advisable to arrange the throttle to shut off completely when it is closed, so that maximum braking effect can be obtained from the engine on the overrun.

3 As an approximate guide, up to $\frac{1}{8}$ throttle is controlled by the pilot jet, from $\frac{1}{8}$ to $\frac{1}{4}$ throttle by the throttle slide cutaway, from $\frac{1}{4}$ to $\frac{3}{4}$ throttle by the needle position and from $\frac{3}{4}$ to full throttle by the size of the main jet. These are only approximate divisions: there is a certain amount of overlap.

4 Note that machines fitted with an air filter have a slightly smaller size of main jet than standard. This is to compensate for the reduced air flow, which tends to make the mixture richer. It follows that when an air filter is disconnected, the larger size main jet must be fitted. If it is not, a weak mixture will result with the possibility of engine damage through overheating.

8.3 Gauze elements can be removed and cleaned in petrol

8 Air filter element: location, examination and cleaning

1 A number of different types of air filters were used during the production run of the Terrier and Tiger Cub models, the type being dependent on the year of manufacture, model and application. The element will be of the gauze or paper type, and may be mounted remotely or attached directly to the carburettor.

2 The remotely mounted types are usually incorporated in or mounted behind the toolbox, and are connected by a rubber pipe to the carburettor mouth. Maintenance of the paper-type element is confined to removing and blowing it clean with compressed air. After it has been cleaned about six times, it is advisable to fit a new replacement.

3 The gauze type elements can be cleaned in petrol and then reused indefinitely. Allow the petrol to evaporate, and oil the gauze before the element is refitted.

4 If the air filter is detached for any reason, it is imperative that the carburettor is re-jetted to compensate for the changes in carburation that will occur. This means INCREASING the size of the main jet by at least 10; a machine fitted with an air filter has a smaller than standard main jet fitted, to compensate for the restriction of air flow imposed by the air filter and the consequent enriching of the mixture.

5 The air filter hose (if fitted) must be in good condition if the filter is to function effectivel;y. Replace any hose that shows signs of either cracking or splitting

9 Exhaust system

1 The exhaust system rarely requires attention because it does not need a regular clean-out like that of a two-stroke. It is important that both the exhaust pipe and silencer are rigidly mounted so that they cannot work loose, and that there is no air leak at any point in the system. An air leak will give rise to an elusive backfire when the engine is on the over-run and can prove difficult to trace.

2 If possible, when renewing the silencer, fit a genuine Triumph part to retain the original design characteristics. Failing that, a good pattern part must be fitted. Especially on non-competition models avoid fitting unsuitable silencers as the change in design will inevitably alter the breathing of the engine, requiring an alteration in main jet sizes and also the throttle slide. Arriving at the correct carburettor settings to meet the requirements of all altered exhaust system requires skill and experimentation that is generally outside the scope of the average owner.

10 Checking the lubrication system

1 The oil return to the tank should be checked as soon as the engine is started. Remove the oil filler cap, and observe the returning stream of oil from the projecting pipe at the base of the neck. This stream of oil will be continuous at first as the scavenge side of the pump picks up the small accumulation of oil which will have drained down into the crankcase. After a few minutes the return will become spasmodic containing as much air as oil. This is because the scavenge system is intentionally of greater capacity than the delivery system, to prevent a build-up of oil in the crankcase.

2 If no return feed is evident, the engine must be stopped and the problem rectified before proceeding any further. Possible causes are wear in the plain timing side bearing (where fitted) causing loss of pressure, leaking or sticking ball valves in the pump, or a leak between the pump mounting face and crankcase.

3 Do not omit to change the engine oil at the recommended intervals. Whilst the tank is drained, the opportunity should also be taken to clean the tank and crankcase filters by washing them in petrol.

11 Oil pump: general description

1   The oil pump on all models is of the twin plunger type, and is driven from the distributor drive shaft. The pump consists of a phosphor-bronze body, in which two parallel bores of different sizes are machined. Two plungers of corresponding size are fitted to the bores, the pair being driven by a common connecting link from an eccentric pin at the bottom of the distributor drive shaft.

2   Each bore and plunger is controlled by two sprung ball valves, which serve to distribute the oil through the relevant port in the casing. Two steel balls of $\frac{3}{16}$ in diameter, followed by a light spring, are fitted to the base of the pump body and are each secured by a threaded hollow plug. A similar pair of valves, using 5/32 in steel balls, are fitted into recesses in the crankcase.

3   It should be noted that on certain late models the oil pump drive was modified. The driving crank was abandoned in favour of a duralumin block which was designed to run in forked ends of the plungers. The two types of pump appear very similar, but are not interchangeable due to machining differences in the crankcase.

11.1 Component parts of the twin plunger oil pump

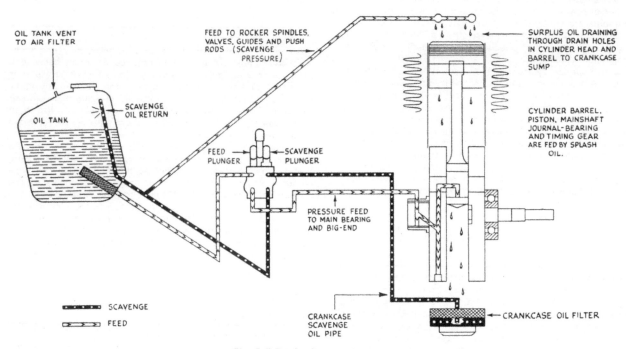

Fig. 2.5 Engine lubrication system

12 Oil pump: removal, examination and maintenance

1   The oil pump is mounted on the right hand crankcase half, and access to it can be gained after removing the right-hand outer and inner covers. On non-distributor models it is necessary to remove the contact breaker plate and ATU, and consequently the ignition timing will be lost and must be set on reassembly.

2   Release the gearchange and kickstart pedals and release the outer cover retaining screw, lifting the cover away. Take care not to damage the gasket, which may be re-used if unbroken. Note that there are two locating dowels which may be displaced. Disengage the kickstart return spring, and remove it, complete with the guide plate. Pull off the felt sealing ring on the gearchange spindle. Remove the gearbox drain plug and allow the oil to drain off into a suitable container.

3   Release the two screws which retain the selector quadrant cover, and remove the cover to give access to the pivot pin. Using a pair of pointed nosed pliers, remove the split pin and then displace and remove the pivot pin. The quadrant will remain in position and need not be disturbed.

4   Release the inner screw retaining screws, then carefully draw the cover off, taking care that the camshaft remains in place in the crankcase. If this precaution is not observed, the tappets and pushrods will become displaced, causing much unnecessary work.

5   Before removing the pump, obtain two small tins into which the balls and spring can be placed. Although the four springs are identical, there are two sizes of ball used, and these must be replaced in their relevant valve seats. Release the two pump mounting bolts, and lift the pump body away, noting the two $\frac{5}{32}$ in steel balls and their springs will drop out of the crankcase. As the pump body is released, disengage the crank or block, depending upon the type fitted.

6   The pump must be kept scrupulously clean during dismantling and reassembly, as even a small piece of fibre from a cleaning rag can impair the operation of one of the ball valves. It is recommended that dismantling is undertaken on a clean piece of cloth or paper. Have a small dish of petrol to hand to wash off each component as it is removed.

7   Dismantling is very straightforward. The plungers can simply be withdrawn from the body and placed to one side. If necessary, remove the small circlip from the clevis pin which retains the plunger ends to the crank, allowing the plungers to be detached. Unscrew the end plugs and remove the two balls and springs.

8   Examine the pump bores and plungers for signs of scoring or other damage. The pump does not normally wear badly, but if the plungers are a sloppy fit or if damage is found, it will be necessary to renew the plungers and body as an assembly. Obtaining a new pump may prove difficult, but it is often possible to obtain a serviceable secondhand item. Examine the steel balls, rejecting any which have become marked or pitted. The conical valve seat should be clear of any obstruction.

9   The sealing face between the valve seats and bores can be restored by placing each ball in position, in turn, and tapping it against its seat, using a small parallel punch and hammer. This ensures that the seat conforms exactly to the profile of the ball. Examine the springs and reject any which appear worn or damaged. It is a good idea to renew the springs as a matter of course, assuming that new components can be obtained.

10   Reassembly is a reversal of the dismantling sequence, ensuring that all parts are clean and well lubricated. Fit the $\frac{3}{16}$ in steel balls in the base of the pump, then reassemble the springs amd plugs. The pump should be refitted to the casing using a new gasket, having first positioned the springs and $\frac{5}{32}$ in steel balls. Fit and tighten the securing bolts.

13 Rocker oil feed: examination and renovation

1   The rocker oil feed is straightforward and is normally reliable if assembled properly, but can give rise to oil leakage problems if mistreated. In the event of leaks occurring the following points should be checked.
      Make sure that the rubber connecting pipe at the tank stub is in sound condition and is securely attached. A couple of turns of soft iron locking wire can be used to secure the tubing, if desired.

2   The copper feed pipe should not pass so close to the engine that vibration causes it to chafe through. If necessary, it can be re-routed by judicious bending. The most common area for leakage to develop is in the vicinity of the banjo unions. Very often the pipe becomes fractured due to the unions twisting when the domed retaining nuts are fitted. This can be avoided by holding the union with pliers whilst the nut is tightened.

3   The pipe can be remade, using the existing unions and T piece. Start by removing the pipe completely and degreasing it . Once clean, a blowlamp can be used to de-solder the unions and T piece. Clean these parts carefully, paying particular attention to the bores in which the pipes fit. The unions can be refaced if marked or damaged , by rubbing them on a piece of fine emery cloth placed over a sheet of glass.

4   Make up the new pipes from copper tubing, which can be obtained from plumbers or central heating engineers. Clean the ends to be soldered, then tin them with solder before the unions and T piece are sweated on. The soldered joint must be even and continuous if leaks are to be avoided.

5   When fitting the pipe to the rocker shafts, fit a $\frac{3}{8}$ in copper washer, the union, and then a $\frac{5}{16}$ in copper washer. The washers should preferably be new, but may be re-used if they are annealed first by heating to red heat, then quenching it in cold water.

6   If pipe fracture proves to be a persistent problem, this can often be cured by fitting a short length of reinforced plastic or synthetic rubber tubing after cutting out a small section of pipe in the vicinity of the fracture. This will absorb and damp vibration, which can cause copper pipe to work harden and crack. Plastic tubing also makes it possible to check the rocker feed visually.

12.7 End plugs house non-return ball valves

12.9 Valve seats can be restored by tapping ball using punch

14 Fault diagnosis: fuel system, carburation and engine lubrication

Symptom	Cause	Remedy
Engine 'fades' and eventually stops	Blocked air hole in filler cap	Clean.
Engine difficult to start. Fuel drips from carburettor	Carburettor flooding	Dismantle and clean carburettor. Check for punctured float.
Engine runs badly. Black smoke from exhaust	Carburettor flooding	Dismantle and clean carburettor. Check for punctured float.
Engine difficult to start. Fires only occasionally and spits back through carburettor	Weak mixture	Check for fuel in float chamber and whether air slide working.
Oil does not return to oil tank	Scavenge system ineffective	Stop engine immediately. Check pump for leaks or sticking valves.
Engine joints leak oil badly	See above	Dismantle and clean pump. Check that scavenge system is operating correctly.
Oil consumption heavy. Blue smoke from exhaust.	Engine in need of rebore	Rebore cylinder and fit O/S piston.

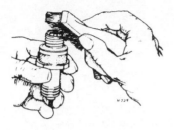

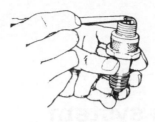

FIG. 5.2. SPARKING PLUG MAINTENANCE

Cleaning deposits from electrodes and surrounding area using a fine wire brush

Checking plug gap with feeler gauges

Altering the plug gap. Note use of correct tool

White deposits and damaged porcelain insulation indicating overheating

Broken porcelain insulation due to bent central electrode

Electrodes burnt away due to wrong heat value or chronic pre-ignition (pinking)

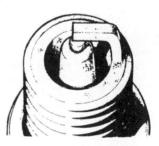

Excessive black deposits caused by over-rich mixture or wrong heat value

Mild white deposits and electrode burnt indicating too weak a fuel mixture

Plug in sound condition with light greyish brown deposits

Sparking plug electrode conditions

Chapter 3 Ignition system

Contents

General description 1
Contact breaker assembly: adjustment 2
Contact breaker assembly: removal, examination and
renovation 3
Automatic timing unit: examination and renovation 4
Condenser: removal and replacement 5

Distributor: examination and renovation 6
Ignition timing adjustment 7
Ignition coil: checking 8
Sparking plug: checking and resetting the gap 9
Fault diagnosis: Ignition system 10

Specifications

Contact breaker gap .... 0·014 – 0·016 in (0·35 – 0·40 mm)

Sparking plug gap .... 0·020 in (0·50 mm)

Sparking plug type:
    T.20S model .... Champion L–5 or Motorcraft AE3
    All other models .... Champion L–7 or L86 or Motorcraft AE3

Ignition timing

Model	Crankshaft degrees	Piston position
T.15 Terrier	8° BTDC	$\frac{1}{64}$ in BTDC
T.20 to 17388	8° BTDC	$\frac{1}{64}$ in BTDC
T.20 from 17388	4° BTDC	TDC
T.20C	4° BTDC	TDC
T.20T	8° BTDC	$\frac{1}{64}$ in BTDC
T.20S	16° BTDC	$\frac{1}{16}$ in BTDC
T.20SS	20° BTDC	0·096 in BTDC
T.20SH	16° BTDC	$\frac{1}{16}$ in BTDC

Note: Timing will vary where different combinations of camshaft, piston and carburettor have been used.

Energy transfer ignition
Alternator rotor timing:
T.20T (Low compression, standard camshaft) .... Rotor keyway No. 2*
T.20S (High compression, high-lift camshaft) .... Rotor keyway No. 1*
* See Fig. 3.1
Contact breaker opens at:

T.20T
8° (0·016 in, 0·4 mm) BTDC

T.20S
16° (0·063 in, 1·6 mm) BTDC

1 General description

On all models up to September 1963, Engine No 94599, a distributor and coil ignition system was used. When the ignition is switched on, current is fed from the battery to the ignition coil primary, or low tension, circuit. As the engine is cranked over, the contact breaker points separate at a predetermined setting, thus interrupting the flow of current through the primary circuit. At this point, high tension voltage is induced in the secondary windings of the ignition coil and this passes to the sparking plug electrodes producing the spark required to ignite the mixture.

There is also an emergency starting position (marked 'EMG') on the ignition switch, to enable the machine to be started with a flat battery. With this position selected, and the contact breaker points closed, the return circuit to the alternator is via one arm of the rectifier bridge. As the points separate, a pulse of electrical energy is discharged through the battery and the primary windings of the ignition coil, causing the induction of a HT spark in the normal way.

As the engine runs, the battery begins to charge, and after a while, the rising battery voltage opposes that of the alternator, thus reducing the energy available for transfer to the coil. In this way, the resulting misfire serves to remind the rider to switch from the 'EMG' position to 'IGN' to resume normal running.

If for any reason the battery cannot be used, it is possible to run the machine with the battery disconnected, provided that the negative (-) battery lead is earthed. It should be noted that the lights will not function until a battery is refitted, and that the machine should not normally be run in this condition.

Later machines have a system which is essentially the same as that described in the foregoing paragraphs, differing only in the respect that the contact breaker and ATU are housed in the right-hand casing, rather than in the distributor.

Energy-transfer ignition system

The above system was fitted to certain competition machines, to enable them to be run without a battery and/or lights during competition events. A specially-wound alternator stator and coil are used, in conjunction with a contact breaker assembly designed to open for a very short interval. The alternator output runs directly to earth until the contacts open. At this point, the current flows to the ignition coil primary windings, inducing the HT spark in the normal way.

As the alternator output is alternating current (ac) it follows that the rotor and crankshaft are arranged so that the pulse is at its highest point as the contacts separate. In view of the importance of obtaining a satisfactory spark at cranking speed, it is essential that this rotor timing is correct. The rotor will be found to have two keyways, these corresponding with low and high compression engines, to cater for the varying characteristics.

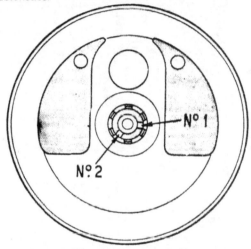

Fig. 3.1 Alternative rotor keyway positions

No 1: High compression, high-lift camshaft models
No 2: Low compression, standard camshaft models

2   Contact breaker assembly : adjustment

1   Remove the distributor cap or contact breaker cover, depending on model, and examine the faces of the contact breaker points. If they appear burnt or pitted, they should be removed for renovation or renewal. See Section 3 for details. If only slightly soiled, they may be dressed in place, using a magneto file or fine emery paper wrapped round a strip of thin sheet metal. Remove any accumulations of dirt using lint-free rag moistened with petrol.
2   Gradually turn the engine over until the contacts can be seen to be fully open. Check the gap, which must be between 0.014 and 0.016 in (0.35 and 0.40 mm), using a feeler gauge. The gauge should be a light sliding fit between the contact faces. If necessary, slacken the fixed contact retaining screw *just* enough to allow it to be moved. The desired setting can now be obtained, and the securing screw tightened, not forgett-

ing to re-check the setting once this has been done. Before refitting the cover, apply a trace of grease or engine oil to the contact breaker cam.

3   Contact breaker assembly: removal, examination and renovation

1   If the points appear badly pitted or worn, they should be removed for further examination. Start by disconnecting the contact breaker lead from the moving contact spring blade, making a careful note of the disposition of the washers to aid reassembly. Slacken and remove the fixed contact retaining screw to allow the assembly to be lifted clear of the baseplate. The moving contact can be lifted off the fixed contact pivot pin to permit close examination of the contact faces.
2   If the contact faces appear badly burnt or pitted, it will be necessary to renew the set. Note that burnt contact faces are quite often due to a faulty condenser allowing arcing to occur as the points separate. See Section 5 for further details. If only lightly marked, it is possible to dress the contact faces using fine emery cloth or an oilstone. The contact faces must be kept square, the object being to remove any material which may have caused localised contact. Where a pip on one contact, and a corresponding pit on the other contact has developed, it is more important to remove the projecting material than to remove the pitting. Finish off with fine polishing paper to produce a bright surface finish. It is considered best to finish the contacts to a slightly convex profile.
3   Before reassembling and resetting the contact breaker assembly, check that the automatic timing unit (ATU) is functioning correctly. Apply a smear of grease or engine oil to the cam face .

4   Automatic timing unit: examination and renovation

1   The ignition timing on post-distributor models Is governed by an automatic timing unit (ATU) mounted behind the contact breaker base plate. As the engine speed rises, centrifugal force causes two bobweights to be thrown outwards which turns the operating cam in relation to the crankshaft. This enables the ignition spark to occur at an earlier point in the ignition cycle.
2   The ATU is retained on a taper by a central retaining bolt, and can be removed after the contact breaker baseplate has been detached. Slacken and remove the central retaining bolt. In the absence of the Triumph extractor, No D485, it is possible to fabricate a suitable tool using two bolts; obtain a bolt of similar dimensions to those of the retaining bolt, then cut if off so that when it is screwed into the threaded camshaft end, the top is $\frac{3}{8}$ - $\frac{1}{2}$ in from the end of the unit. Cut a screwdriver slot in the end, using a hacksaw, then fit it in position. Obtain a second bolt which will screw into the larger outer thread. Thread this into the end of the unit so that It bears upon the head of the smaller bolt. Tighten the bolt so that the assembly is under tension, at which point, a light tap on the head of the larger bolt should dislodge the ATU.
3   Examine the unit for wear, checking that the bobweights and cam move easily. If the weights seem sloppy, check that the pivot pins have not worked loose in the baseplate. If this is the case, it is possible to re-rivet them using a hammer and punch, taking care not to distort the pins or the base plate. If the pivot holes in the bobweights have become elongated, or the pivot pins have worn, it will be necessary to renew the unit, otherwise the ignition timing will become erratic.
4   Check that the springs have not become stretched, which would result in the ignition advancing too early. The contact breaker cam is not normally subject to very much wear, and periodic lubrication with a trace of grease is normally sufficient to ensure that it lasts at least as long as the rest of the unit. If a new unit is required, it is important to specify the model, year and engine number, so that the correct unit is supplied.

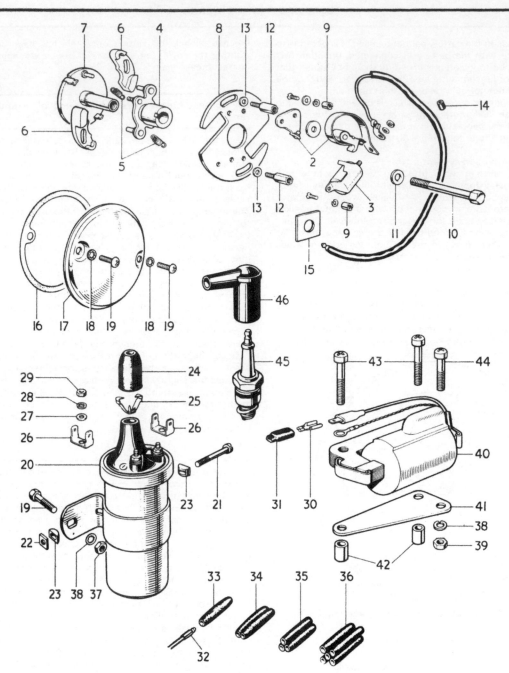

Fig. 3.2 Contact breaker assembly and ignition coils

1 Contact breaker assembly	17 Cover	33 Single snap connector – as required
2 Contact set	18 Serrated washer – 2 off	34 Double snap connector – as required
3 Condenser	19 Bolt – 2 off	35 Triple snap connector – as required
4 Cam	20 Ignition coil	36 Quintruple snap connector – as required
5 Spring set	21 Screw	
6 Weight – 2 off	22 Square nut	37 Nut – 2 off
7 Sleeve and action plate	23 'D' washer – 2 off	38 Serrated washer
8 Base plate	24 Grommet	39 Nut
9 Nut – 2 off	25 Clip	40 Ignition coil (alternative type)
10 Bolt	26 Lucar terminal	41 Bracket
11 Washer	27 Washer – 2 off	42 Spacer – 2 off
12 Bolt – 2 off	28 Serrated washer – 2 off	43 Screw – 2 off
13 Washer – 2 off	29 Nut – 2 off	44 Screw
14 Grommet	30 Lucar connector – as required	45 Sparking plug
15 Cable clip	31 Insulating cover – as required	46 Plug cover
16 Gasket	32 Snap connector terminal – as required	

4.2a Contact breaker baseplate must be removed ...

4.2b ... to gain access to automatic timing unit

4.3 Check the ATU for wear and slack operation

5 Condenser: removal and replacement

1 A condenser is included in the contact breaker circuitry to prevent arcing across the contact breaker points as they separate. It is connected in parallel with the points and if a fault develops the ignition system will not function correctly.

2 If the engine is difficult to start or if misfiring occurs, it is possible that the condenser has failed. To check, separate the contact breaker points by hand whilst the ignition is switched on. If a spark occurs across the points and they have a blackened or burnt appearance, the condenser can be regarded as unserviceable.

3 It is not possible to check the condenser without the necessary test equipment. In view of the low cost of a replacement it is preferable to change the condenser and observe the effect on engine performance.

4 To remove the condenser, unscrew the terminal nut on the end and lift off the contact breaker return spring. Remove the retaining screw and lift out the condenser.

6 Distributor: examination and renovation

1 Terrier, and Tiger Cub models prior to September 1963, were equipped with a distributor driven from the crankshaft, and mounted vertically on the right-hand side of the engine unit. The distributor was retained, on pre-1959 models by a small clamp at the base of the unit. Subsequent models differed slightly, in that the clamp was relocated inside the outer casing, the screw head projecting through the side of the case.

2 Generally speaking, the remarks on servicing given in Sections 2,3 and 4 can be applied to the distributor, although it should be noted that the ATU forms an integral part of the distributor shaft. Maintenance is normally confined to applying a smear of grease to the contact breaker cam, and a few drops of light machine oil to the moving parts of the advance mechanism. Take care not to contaminate the contact breaker points, as oil will form an effective insulating film.

3 Keep the distributor body and cover clean and dry, and remove any surface grime from the inside of the cap. In the event of misfiring problems in wet weather, a proprietary de-watering aerosol spray may be used to prevent tracking to earth.

4 If necessary, the distributor may be removed for dismantling after releasing the clamp. It should be noted that it is not practicable to renew the distributor body bearings, and if worn, a new or good secondhand replacement should be fitted.

5 When refitting the distributor assembly, check that the 'O' ring is in sound condition. Set the engine at TDC on the compression stroke. Turn the distributor shaft anticlockwise so that the contact breaker is in the fully open position. Offer up the distributor, noting that the assembly should be at the seven o'clock position, assuming that the operator is facing the timing side of the engine. The contact breaker gap should be checked and reset, and the ignition system re-timed, as described in the following section.

7 Ignition timing adjustment

A. Distributor ignition models (all models up to September 1963)

1 The ignition timing may be set using either a degree disc bolted to the crankshaft, or a plunger arrangement in the sparking plug hole. It will be necessary to arrange the machine so that the rear wheel is supported clear of the ground and is free to rotate. Remove the sparking plug and select top gear. It is

now necessary to establish the position at which the piston is at top dead centre (TDC) on the compression stroke.

2   If a degree disc is to be used, a stop which will screw into the sparking plug hole to stop the piston part way up the bore will be required. A suitable stop may be fabricated using an old sparking plug: clamp the plug in a vice, and proceed to hacksaw or file off the topmost ridge of the metal body. This ridge is spun over to retain the insulator and electrode assembly, and the latter can be removed and discarded after the ridge has been cut off. The threaded metal portion of the plug will be left, and the earth electrode must be removed from this. Cut a slot scross the threads to allow the compression pressure to escape.

3   Obtain a bolt of about 3 in length which will fit into the body. The diameter is not critical. Assemble the bolt, retaining it with suitable washers and a nut, so that the main part of the thread projects out of the bottom of the plug body.

4   Turn the engine over, by way of the rear wheel, until the inlet valve closes and the piston begins to ascend on the compression stroke. Fit the stop into the sparking plug hole, and continue turning the wheel until the piston comes up against the end of the stop. Fit the timing disc to the end of the crankshaft, having first removed the chaincase, then affix a suitable wire pointer to the crankcase to align with 0° on the timing disc.

5   Turn the engine back until the stop is again reached, and note the number of degrees covered. This figure should be halved, and the engine turned gradually *forward* by that many degrees. Reset the timing disc to read 0° of that point, which will be TDC compression . The stop may now be removed. (For example, if the total movement between stops is 86°, the engine should be turned forward by 43°, and the timing disc set at zero).

6   Having established the exact position of TDC, remove the distributor cover and check that the contact breaker points are clean and correctly adjusted. Make up a simple test apparatus, by connecting a torch battery and bulb as described in the next paragraph.

7   Solder a lead between one of the battery terminals to the base of a suitably rated bulb. A second battery lead should be affixed to the other terminal and the free end be attached to a crocodile clip. Connect a third lead to the side of the bulb, and fix to the free end, a second crocodile clip. It will be found that the bulb will light when contact is made between the two crocodile clips.

(NOTE: Owners possessing a multimeter may wish to use this in place of the battery/bulb arrangement).

8   Turn the engine backwards by about 20°, and connect one of the crocodile clips to the contact breaker terminal, and the second clip to earth. Now turn the engine very slowly forwards until the light *just* goes out. Note the reading at this point, which should correspond with the following figures:

Model	Ignition advance
T15 (Terrier)	8° BTDC
T20 (up to engine no: 17388)	8° BTDC
T20 (from 17389 to 35846)	4° BTDC
T20 (from 35847 on)	4° BTDC
T20C (all models)	4° BTDC
T20T (all models)	8° BTDC
T20S (all models)	16° BTDC

9   If the reading indicated differs from that given above, slacken the distributor clamping screw, set the crankshaft at the right position, then move the distributor body to and fro until the light just goes out. Retighten the screw, turn the engine over once or twice, then re-check the timing.

10   The ignition timing may also be set using a suitably mounted dial gauge, set so that the pointer passes down through the sparking plug hole and touches the piston crown. Find TDC compression in the same way as described for the degree disc method; set up the dial gauge and measure the maximum point of piston travel, then re-set the gauge to zero.

Back off the piston, connect the test lamp arrangement, and measure the position BTDC at which the lamp is extinguished. This should be adjusted, if necessary, to conform to the following positions:

Piston position

T15 (Terrier - all models):	$\frac{1}{64}$ in BTDC
T20 (up to 17388):	$\frac{1}{64}$ in BTDC
T20 (17388 on):	at TDC
T20C (all models):	at TDC
T2OT (all models):	$\frac{1}{64}$ in BTDC
T20S (all models):	$\frac{1}{16}$ in BTDC

11   Adjustment is accomplished as described earlier, by slackening the clamp screw of the distributor and moving the body, before retightening when it is in the correct position.

B. Non-distributor ignition models (September 1963 on)

The distributor arrangement was eventually abandoned in favour of a system in which the contact breaker assembly and automatic timing unit were incorporated in the right-hand outer cover. An inspection cover is provided, retained by two screws.

12   Timing checking and adjustment is carried out in the same way as that described for the distributor models, using a battery/bulb apparatus in conjunction with a degree disc or a dial gauge. The static timing setting should be as follows:

Model	Degrees	Inches	mm
T20	8°	0.016	0.40
T20SS	20°	0.096	2.45
T20SH	16°	0.060	1.5

13   If necessary, the timing may be adjusted as follows; set the crankshaft at the correct position BTDC. Slacken the contact breaker baseplate sleeve nuts sufficiently to allow the plate to be moved. Set the plate so that the contact breaker points *just* separate (test lamp will go out) then tighten the retaining nuts. Turn the engine over a few times and re-check the setting before refitting the outer cover.

C. Ignition timing: general notes

14   To obviate the need to employ the lengthy procedures described in the foregoing sections, it is worthwhile devising a set of alignment reference marks during the initial setting up of the ignition timing, particularly if it is intended to keep the machine for any length of time.

15   Obtain a few inches of fairly robust steel strip, and from it fabricate a pointer. The pointer should be arranged so that it runs close to the edge of the alternator rotor. Remove one of the crankcase bolts from inside the chain case, and fit the pointer over the bolt. The bolt may now be refitted and tightened, ensuring that the business end is positioned about $\frac{1}{32}$ in away from the rotor face. See Fig. 3.4.

Using the piston stop method, or a dial gauge, as described earlier, find the position of TDC compression. Make sure that this position is established accurately. Using a sharp scribing tool, mark the rotor with a line corresponding to the end of the pointer. The line end should be marked: 'TDC'. Now establish the correct ignition advance position for the relevant model, and mark a second line, which should be captioned IGN or T to identify it as the ignition timing mark. These latter marks may then be used each time the ignition timing is checked. Note that if the alternator rotor is removed, or if the distributor or ATU is disturbed, the full procedure should be undertaken to ensure no error has crept in due to these components being moved.

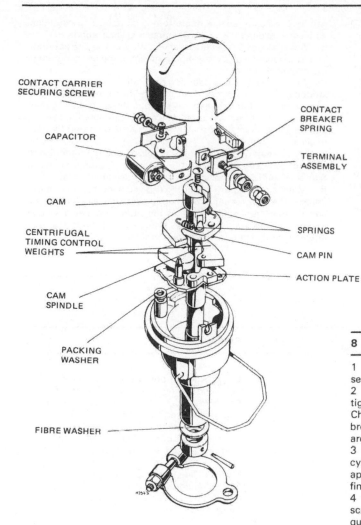

CONTACT CARRIER
SECURING SCREW

CAPACITOR

CAM

CENTRIFUGAL
TIMING CONTROL
WEIGHTS

CAM
SPINDLE

PACKING
WASHER

FIBRE WASHER

CONTACT
BREAKER
SPRING

TERMINAL
ASSEMBLY

SPRINGS

CAM PIN

ACTION PLATE

Fig. 3.3 Distributor – component parts

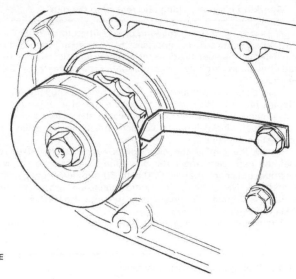

Fig. 3.4 Timing pointer arrangement (see text)

8   Ignition coil: checking

1   The ignition coil is a sealed unit, designed to give long service. It is mounted on the frame, beneath the dualseat.

2   To test the coil, first ensure that the terminals are clean and tight and the HT lead is correctly connected within the coil. Check that the battery is fully charged and remove the contact breaker dust cover, turn the engine over slowly until the points are closed.

3   Wedge the sparking plug end of the HT lead between two cylinder barrel fins so that the bared end of the centre wire is approximately $\frac{3}{16}$ in - $\frac{1}{4}$ in away from the metal of the cooling fins.

4   Switch on the ignition and, using a plastic handled screwdriver flip the contact breaker points open - a healthy and quite audible spark should jump from the end of the HT lead to the cylinder barrel fins. Repeat the operation several times, then switch off the ignition to avoid damage to the coil.

5   If no spark results and it is known that the battery, condenser and contact breaker points are not at fault, take it to an electrical repair expert for checking. A faulty coil must be renewed as it is not practicable to effect a repair.

7.12 Multimeter can be used to check ignition timing

8.1 Ignition coil is normally located beneath dualseat

9 Sparking plug: checking and resetting the gap

1 A 14 mm, $\frac{1}{2}$ in reach sparking plug is fitted to all Tiger Cub and Terrier models. It is important that a sparking plug of the type listed in the specifications, or an equivalent, is used, as the grade has been carefully chosen to match the characteristics of the engine.

2 It is particularly important that a sparking plug having a different reach is not used, as there is a risk of a longer reach plug striking the piston crown. Similarly, the electrodes of a shorter reach plug will not be correctly positioned for optimum combustion efficiency.

3 Check the gap at the plug points every 2,000 miles. To reset the gap, bend the outer electrode closer to the central electrode and check that a 0.020 in (0.50 mm) feeler gauge can be inserted. Never bend the central electrode, otherwise the insulator will crack, causing engine damage if particles fall in whilst the engine is running.

4 The condition of the sparking plug electrodes and insulator can be used as a reliable guide to engine operating conditions, with some experience. See accompanying illustrations.

5 Always carry one spare sparking plug of the correct grade. This will serve as a get-you-home means if the sparking plug in the engine should fail.

6 Never overtighten a sparking plug, otherwise there is risk of stripping the threads from the cylinder head, especially in the case of one cast in light alloy. A stripped thread can be repaired by using what is known as a 'Helicoil' thread insert, a low cost service of cylinder head reclamation that is operated by many dealers.

7 Use a sparking plug spanner that is a good fit, otherwise the spanner may slip and break the insulator. The plug should be tightened sufficiently to seat firmly on its sealing washer.

8 Make sure the plug insulating cap is a good fit and free from cracks. The cap contains the suppressor that eliminates radio and TV interference; in rare cases the suppressor has developed a very high resistance as it has aged, cutting down the spark intensity and giving rise to ignition problems.

10 Fault diagnosis: ignition system

Symptom	Cause	Remedy
Engine will not start	No spark at plug	Check whether points open and close. Check also whether points are dirty – if so, clean. Check whether points arc when engine is turned over. If so, condenser has failed. Check that low tension circuit is connected properly. Check battery connections and condition.
Engine starts, but runs erratically	Intermittent or weak spark	Try renewing sparking plug. Check ignition timing. Check plug lead and low tension wiring for short circuits.
Engine will not run at low speeds, kicks back during starting	Ignition over-advanced	Re-set ignition timing. Check operation of ATU
Engine lacks power, overheats	Ignition timing retarded	Re-set ignition timing. Check operation of ATU.
Engine misfires at high speeds	Incorrect sparking plug	Check plug grade with list of recommendations.

Chapter 4 Frame and Forks

Contents

General description 1	Swinging arm rear suspension: examination and renovation 14
Front fork legs: removal from frame 2	Rear suspension units: examination 15
Steering head assembly: removal from frame 3	Centre stand: examination 16
Lower legs: removal with forks installed in frame 4	Prop stand: examination 17
Front forks: dismantling 5	Footrests: examination and renovation 18
Front fork oil seals: examination and renewal 6	Speedometer: removal and replacement 19
Front forks: examination and replacement of bushes 7	Speedometer cable: examination and renovation 20
Front forks: examination and renovation – general 8	Tachometer: removal and replacement 21
Steering head bearings: examination and replacement . . . 9	Tachometer drive cable: examination and renovation 22
Front forks: reassembly 10	Tank badges and knee pads 23
Front forks: damping action 11	Steering head lock 24
Frame assembly: examination and renovation 12	Cleaning: general 25
Plunger rear suspension: examination and renovation . . 13	Fault diagnosis: frame and forks 26

Specifications

Front forks:

Lightweight telescopic forks, internally sprung. (Most roadster models)
Heavyweight telescopic forks, hydraulically damped. (Sports and competition models)

Damping oil

Type	SAE 30 engine oil or Fork oil
Capacity:	
Lightweight forks	$\frac{1}{8}$ imp pint (75 cc)
Heavyweight forks	$\frac{1}{4}$ imp pint (150 cc)

Frame:

Early models	Plunger type spring frame
Later models	Swinging arm frame with hydraulically damped suspension units

Plunger rear suspension units	
Compression spring free length	$4\frac{7}{8}$ in (124 mm)
Rebound spring free length	$2\frac{3}{4}$ in (70 mm)

1 General description

1   All Terrier, and early Tiger Cub models, employ a spring frame, on which rear suspension is provided by plunger units. Each of these units comprise a plunger with a forked end to accept the wheel spindle, running on a rod clamped vertically to the frame. The movement of the plunger is controlled by a heavy compression spring, and a lighter rebound spring, these being effectively enclosed by a gaiter and shrouds.

2   Later models adopted a swinging arm type frame in which the rear wheel is supported by a pivoted fork, suspended on two hydraulically damped sealed suspension units.

2 Front fork legs: removal from frame

1   It is unlikely that the front forks will need to be removed from the frame as a complete unit unless attention to the steering head bearings is required, or damage to the forks has been sustained as the result of an accident. It is normally more practicable to remove the individual fork legs from the yokes, or if attention to the bushes and seals is required, to remove the lower legs only.

2   Start by placing the machine on the centre stand, if necessary placing wooden blocks beneath the crankcase to raise the front wheel clear of the ground. Release the front

brake cable at the wheel end, by pulling out the split pin, and then displacing the clevis pin from the actuating lever. Release the cable adjuster from the lower fork leg, and lodge the cable away from the fork assembly.

3   Slacken the nuts which retain the wheel spindle clamps, after releasing the torque arm securing bolt (where applicable). The front wheel can now be removed and placed to one side. On machines fitted with a headlamp nacelle, this should be removed as follows; release the headlamp retaining screw from the underside of the nacelle, and lift the headlamp assembly clear. Disconnect the headlamp leads at the various bullet connectors, and place the assembly in a safe place.

4   Release the knurled ring which retains the speedometer drive cable to the underside of the instrument. The upper section of the nacelle is retained by six screws and nuts. Take care not to drop the nuts into the lower section during removal. Lift the top section away, disconnecting the plug-in connectors from the undersides of the switches, or in the case of the combined switch unit, by releasing the switch body after removing the retaining ring. It is now possible to gain access to the fork top nuts. Note that the handle bars should be released by removing the 'U' bolts which retain them, to allow the top nuts to be removed. Remove the two filler plugs which are located a few inches below the tops of the stanchions.

5   On all models, release the nuts which retain the mudguard stays to the lower legs. Remove the mudguard and place it to one side. Slacken the gaiter clips, where fitted. Remove the fork top nuts, and slacken the pinch bolts in the lower fork yoke. Grasp each fork leg in turn and pull it downwards, to clear the yokes. If it proves difficult to dislodge the stanchion from the yokes, temporarily refit the top nut by a few threads, then tap it downwards, using a hide mallet to jar the stanchions free. If necessary, the clamp in the lower yoke may be spread slightly by inserting a screwdriver in the gap, and levering it apart very gently.

3   Steering head assembly : removal from frame

1   It is possible to remove the steering head assembly together with the fork legs, if this is desired. However, it is normally more convenient to remove the fork legs as described in the preceding Section, as this makes removal of the steering head components less unwieldy, and involves very little additional work.

2   On machines fitted with a headlamp nacelle, it is more convenient to release the lower section to give better working clearance. On machines equipped with a conventional headlamp unit, this should be removed after disconnecting the leads at the bullet connectors inside the shell. Place some rag on the top of the petrol tank, then release the handlebars and lay them on the rag.

3   Slacken the pinch bolt on the top fork yoke, and remove the large sleeve nut from the top of the steering stem. Remove the top yoke by knocking it upwards, off the steering column, supporting the lower yoke assembly as this is done. Carefully lower the bottom yoke, catching the steering head balls which will almost certainly be displaced. Note that each race contains fifteen $\frac{1}{4}$ in steel balls.

4   Lower legs : removal with forks installed in frame

1   It is possible to remove the lower legs without disturbing the stanchions, and this method has certain advantages. For example, it is difficult to devise a practical method of filling the lightweight forks on machines fitted with a headlamp nacelle, as the filler plug is small and inaccessible. It is as easy to remove the lower legs, allowing the oil to be emptied, the leg flushed out and the bushes and seal inspected before refilling and fitting the lower legs.

2   With the front wheel removed as described in Section 2, attach a self-locking wrench to the lug at the bottom of the lower leg as a means of holding it whilst the sleeve nut is slackened. The latter can be released using a 'C' spanner, or if not available, a strap wrench. As an alternative, one or two large worm drive hose clips can be clamped around the nut, and the clip boss used as a means of tapping the nut loose with a punch and hammer.

3   With the nut slackened, the lower leg will drop slightly, being retained by the spring which screws into the bottom of the leg. Unscrew the lower leg to release it from the spring, and lift it away. This method may also be applied to the heavyweight forks, the removal procedure being similar.

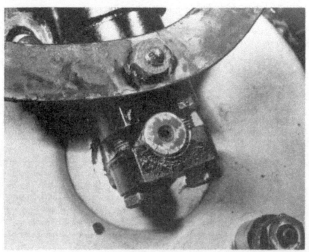

2.3 Release wheel spindle clamps to allow wheel to be removed

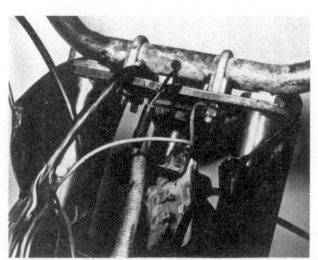

2.4a Release upper nacelle half to gain access to forks

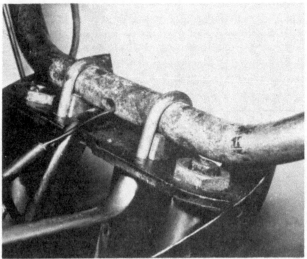

2.4b Handlebars must be removed to gain access ...

2.4c ... to fork top nuts

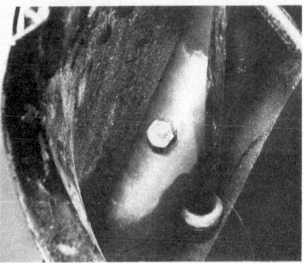

2.4d Remove filler plugs to allow stanchions to pass through yokes

2.5a Slacken and remove bolts which retain upper ...

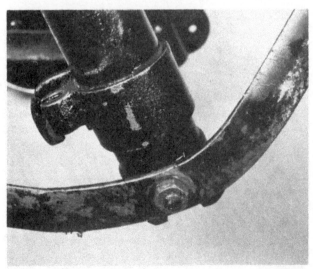

2.5b ... and lower mudguard stays

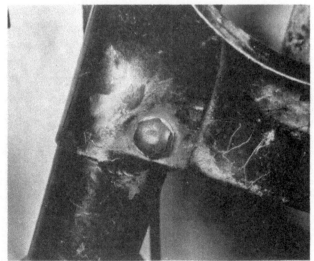

2.5c Slacken the pinch bolt in the lower yoke

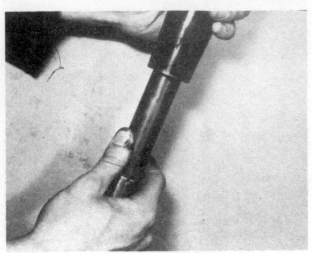

2.5d ... and draw fork assembly downwards

5 Front forks: dismantling

1   Commence by draining the fork legs, if this has not been done already. This is best accomplished by inverting the legs over a suitable drain tray or bowl and allowing the oil to drain out. Pumping the fork will help to speed up this operation. On models fitted with gaiters, these should be removed if still attached to the fork legs. Dismantle each leg individually, so that there is no risk of parts becoming interchanged.

2   Remove the fork spring. On lightweight models fitted with an internal spring, this can be unscrewed from the base of the lower leg and pulled out of the stanchion. On models having an external spring, it can be lifted off.

3   It is now necessary to release the sleeve nut from the top of the lower leg. Prior to this, remove the drain plug screw and the restrictor rod bolt from the base of the lower leg on early heavyweight forks (internal springs); withdraw the restrictor assembly from the top of the stanchion. On the lightweight fork, sleeve nut removal can be accomplished by the use of a 'C' spanner, if available. On the heavyweight fork, and also on the lightweight type if no 'C' spanner is to hand, a strap wrench should be used, with the bottom lug held in a vice. Alternatively, a large worm drive hose clip can be clamped around the nut, the boss being used as a stop with the assembly held in soft vice jaws. A self-grip wrench can be applied to the wheel spindle lug to unscrew the lower leg. Once slackened, the stanchion can be

withdrawn and the seal and bushes checked for wear.

4   It should be noted that a restrictor is mounted at the bottom of the lower leg of later heavyweight forks, being retained by a single bolt which passes up from the underside of the lower leg. It is not normally necessary to remove this component as it is not subject to any significant wear.

6 Front fork oil seals: examination and renewal

1   If signs of oil leakage have been present during use, the seals should be renewed without question. Otherwise, examine the sealing face for signs of wear or damage. This is normally apparent in the form of grooves or scored lines on the seal face, and is caused by a damaged or worn stanchion or a particle of grit. The stanchion must be examined in conjunction with the oil seal, as there is little point in fitting a new seal to a worn stanchion.

2   On the heavyweight fork, the worn seal can be drifted out of the sleeve nut, and a new one fitted in a similar manner. Take great care to fit the new seal squarely into position without damaging the sealing lip.

3   On lightweight forks, the seal takes the form of a square section 'O' ring, and locates in a groove in the sleeve nut. Removal is a simple matter of prising the old seal out of the nut with a small screwdriver, the new seal being pushed into place.

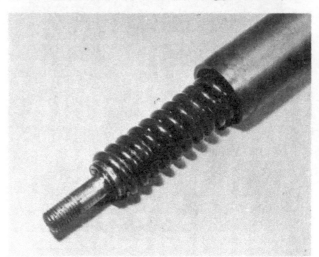

5.2 Remove fork spring from stanchion

5.3a Slacken sleeve nut using C-spanner or strap wrench

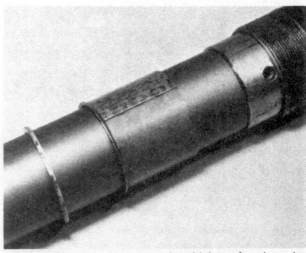

5.3b Stanchion assembly can now be withdrawn from lower leg

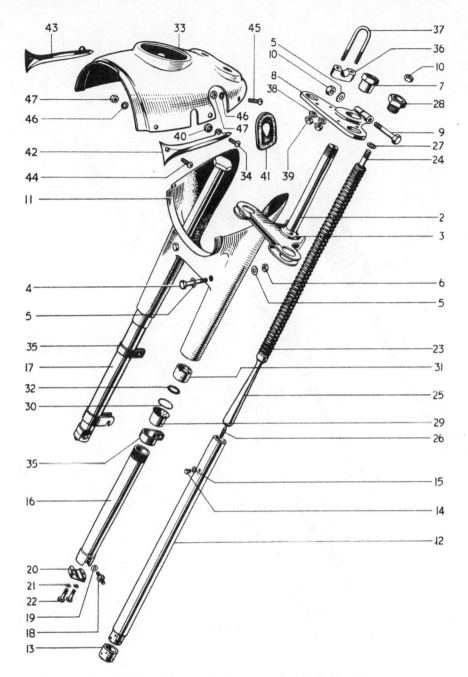

Fig. 4.1 Telescopic front forks – T20 model

1 Fork assembly	12 Stanchion – 2 off	24 Stud – 2 off	36 Handlebar support – 2 off
2 Lower steering yoke and stem	13 Lower bearing – 2 off	25 Restrictor rod – 2 off	37 'U' bolt – 2 off
	14 Filler plug – 2 off	26 Stud – 2 off	38 Washer – 4 off
3 Bottom cone	15 Washer – 2 off	27 Serrated washer – 2 off	39 Nut – 4 off
4 Pinch bolt – 2 off	16 Lower fork leg – LH	28 Cap nut – 2 off	40 Rubber grommet – 3 off
5 Washer – 5 off	17 Lower fork leg – RH	29 Top bearing – 2 off	41 Rubber grommet – 2 off
6 Nut – 2 off	18 Drain stud – 2 off	30 Steel washer – 2 off	42 Motif – LH
7 Sleeve nut	19 Washer – 2 off	31 Dust excluder – 2 off	43 Motif – RH
8 Upper steering yoke	20 Fork end cap – 2 off	32 Oil seal	44 Screw – 3 off
9 Pinch bolt	21 Washer – 4 off	33 Nacelle top	45 Screw – 4 off
10 Nut – 3 off	22 Bolt – 4 off	34 Screw – 2 off	46 Serrated washer – 4 off
11 Lower fork shroud	23 Fork spring – 2 off	35 Mudguard clip – 2 off	47 Nut – 4 off

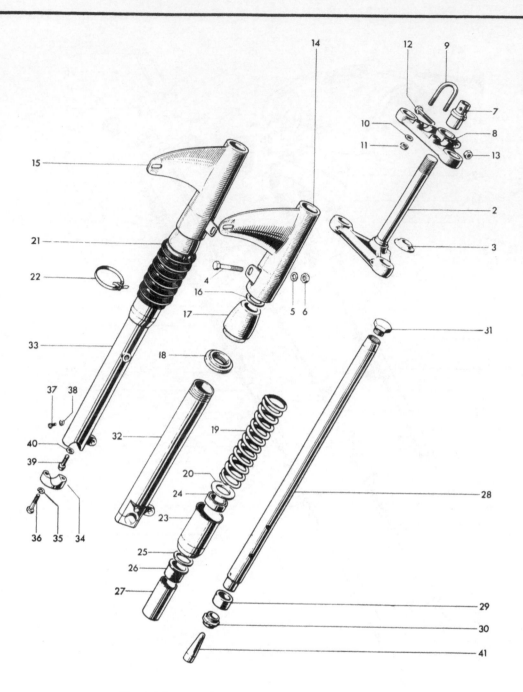

Fig. 4.2 Telescopic front forks – later 'heavyweight' type

1 Fork assembly
2 Lower steering yoke and stem
3 Bottom cone
4 Pinch bolt – 2 off
5 Washer – 2 off
6 Nut – 2 off
7 Fork stem sleeve nut
8 Upper steering yoke
9 'U' bolt – 2 off
10 Washer – 4 off
11 Nut – 4 off
12 Pinch bolt
13 Nut

14 Fork shroud – LH
15 Fork shroud – RH
16 Cork washer – 2 off
17 Upper spring seat – 2 off
18 Lower spring seat – 2 off
19 Fork spring – 2 off
20 Washer – 2 off
21 Telescopic gaiter – 2 off
22 Gaiter clip – 4 off
23 Sleeve nut – 2 off
24 Oil seal – 2 off
25 Washer – 2 off
26 Upper bearing – 2 off
27 Bush – 2 off

28 Stanchion – 2 off
29 Lower bearing – 2 off
30 Nut – 2 off
31 Cap nut – 2 off
32 Lower fork leg – LH
33 Lower fork leg – RH
34 Fork end cap – 2 off
35 Spring washer – 4 off
36 Bolt – 4 off
37 Drain plug – 2 off
38 Washer – 2 off
39 Bolt – 2 off
40 Aluminium washer – 2 off
41 Restrictor – 2 off

6.1a This seal has been very badly scored by worn stanchion

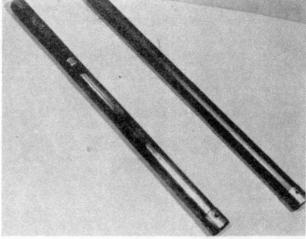

6.1b Comparison of worn (left) stanchion with new component shows severe scoring caused by corrosion

7 Front forks: examination and replacement of bushes

1 Some indication of the extent of wear of the fork bushes can be gained when the forks are being dismantled. Pull each fork stanchion out until it reaches the limit of its extension and check the side play. In this position the two fork bushes are closest together, which will show the amount of play to its maximum. Only a small amount of play that is just perceptible can be tolerated. If the play is greater than this, the bushes are due for replacement.

2 It is possible to check for play in the bushes whilst the forks are still attached to the machine. If the front wheel is gripped between the knees and the handlebars rocked to and fro, the amount of wear will be magnified by the leverage at the handlebar ends. Cross-check by applying the front brake and pushing and pulling the front wheel backwards and forwards. It is important not to confuse any play that is evident with slackness in the steering head bearings, which should be taken up first.

3 The fork bushes can be slid off the fork stanchions if they are clamped in a vice fitted with soft clamps and the large nut on the extreme end removed. If the replacement bushes are a slack fit on the tubes, wear has occurred on the tubes also, in which case a specialist repair with undersize bushes must be made.

4 The fit within the lower fork legs is also important. If wear of the inner surface is evident, it may be necessary to fit lower bushes that have a slightly greater outside diameter.

5 On lightweight forks, the lower bush is retained in a slightly different manner, being staked on to the stanchion. The old bush can be removed by carefully drilling out the countersunk areas. When fitting a new bush, check that the holes align, then carefully stake the bush onto the stanchion with a suitable punch. Take great care to support the bush and stanchion to avoid damage to either part.

8 Front forks: examination and renovation – general

1 Apart from the oil seals and bushes, it is unlikely that the forks will require any additional attention, unless the fork springs are weak or broken. If the fork legs or yokes have been damaged in an accident, it is preferable to have them replaced. Repairs are seldom practicable without the appropriate repair equipment and jigs, furthermore there is also the risk of fatigue failure.

2 Visual examination will show whether either the fork legs or the yokes are bent or distorted. The best check for the fork legs is to remove the fork bushes, as described in Section 7 of this Chapter and roll the legs on a sheet of plate glass. Any deviation from parallel will be immediately obvious.

9 Steering head bearings: examination and replacement

1 Before commencing to reassemble the forks, inspect the steering head races. The ball bearing tracks should be polished and free from indentations and cracks. If signs of wear are evident, the cones and cups must be renewed. They are a tight press fit and must be drifted out of position.

2 Ball bearings are cheap. If there is any reason to suspect the condition of the existing ball bearings, they should be replaced without question. Note that each race contains fifteen $\frac{1}{4}$ in steel balls.

3 Use thick grease to retain the ball bearings in position, whilst the head stem is being assembled and adjusted.

10 Front forks: reassembly

1 To reassemble the forks, follow the dismantling procedure in reverse. Take particular care when passing the sliding fork members through the oil seals, which should be fitted with the lip facing downwards. It is a wise precaution to wind a turn or so of medium twine around the undercut at the base of the thread of the plated collars, to act as an extra seal (heavyweight fork). On early heavyweight forks (internal springs), ensure that the restrictor rod slot is correctly located with the drain plug screw on reassembly.

2 Tighten the steering head carefully, so that all play is eliminated without placing undue stress on the bearings. The adjustment is correct if all play is eliminated and the handlebars will swing to full lock of their own accord when given a light push on one end.

3 It is possible to place several tons pressure quite unwittingly on the steering head bearings, if they are overtightened. The usual symptom of over-tight bearings is a tendency for the machine to roll at low speeds, even though the handlebars may appear to turn quite freely.

4 One problem that will arise during reassembly is the reluctance of the fork main tubes to pass up into the fork top yoke. To facilitate assembly, a broom handle of the correct diameter can be used to good effect, if it is first screwed into the end of the thread of each fork tube. Care should be taken in this instance, to prevent particles of wood from falling into the fork tubes.

5   If, after assembly, it is found that the forks are incorrectly aligned or unduly stiff in action, loosen the front wheel spindle, the two caps at the top of the fork legs and the pinch bolts in both the top and bottom yokes. The forks should then be pumped up and down several times to realign them. Retighten all the nuts and bolts in the same order, finishing with the steering head pinch bolt.

6   This same procedure can be used if the forks are misaligned after an accident. Often the legs will twist within the fork yokes, giving the impression of more serious damage, even though no structural damage has occurred.

7   Do not omit to add the correct amount of damping oil to each fork leg before replacing the fork caps. See Specifications list for the amount and viscosity of oil to be added. Tighten the gaiter retaining clips (where applicable) noting that if the gaiters have air vent holes, they must face downwards to prevent the ingress of water.

10.7 Drain plug doubles as mounting stud for mudguard

11 Front forks: damping action

1   Each fork leg contains a predetermined quantity of oil of recommended viscosity, which is used as a damping medium to control the action of the compression springs within the forks when various road shocks are encountered. If the damping fluid is absent, there is no control over the rebound action of the fork springs and fork movement will be excessive, giving a very 'lively' ride. Damping restricts fork movement on the rebound and is progressive in action - the effect becomes more powerful as the rate of deflection increases.

2   The damping action can be varied only by changing the viscosity of the oil used as the damping medium; it is not practicable to vary the size of the holes in the inner fork tubes. In temperate climates an SAE 30 oil or a proprietary fork oil is used but if considered necessary, the viscosity rating can be increased without any harmful effects.

12 Frame assembly: examination and renovation

1   The frame is unlikely to require attention unless damage has been sustained as the result of an accident. Frame repairs are best entrusted to a specialist in this type of repair work, who will have all the necessary jigs and mandrels available to ensure the correct alignment. In many instances a replacement frame from a breaker's yard is the cheaper and more satisfactory alternative.

2   If the machine is stripped for an overhaul, this affords an excellent opportunity to inspect the frame for signs of cracks or other damage that may have occurred in service. Check the front down tube at the point immediately below the steering head, which is where a break is most likely to occur. Check the top tube of the frame for straightness - that is the tube most likely to bend in the event of an accident.

13 Plunger rear suspension: examination and renovation

1   Early models were equipped with plunger rear suspension in which the frame is supported on two large coil springs. A pair of secondary springs are employed to absorb rebound movement. The fork end is mounted between the two springs, and slides on a rod bolted between the two frame clamps. The assembly is enclosed by a shroud over the upper spring, the lower spring being gaitered.

2   The fork ends are fitted with grease nipples and the manufacturer recommends that the units should be greased every 1000 miles. No other maintenance is normally required. Dismantling becomes necessary only in the event of play developing in the plunger units. Ideally, a Triumph spring compressor, part number Z100, should be to hand during this operation, but dismantling and reassembly are possible without it.

3   Commence by removing the rear wheel, as described in Chapter 5, Section 5, Place some stout wooden blocks beneath the lower frame member so that the rear of the frame is firmly supported and completely stable. Slacken and remove the locknuts on the undersides of the lower mounting lugs, then remove the clamp bolts from the upper lugs. Using a large screwdriver in the slot provided, unscrew the plunger guide rod and pull it upwards, clear of the unit.

4   Some care must be exercised at this stage, as the plunger assembly is free to be displaced, and is under spring tension. It is a wise precaution to place some heavy fabric over the unit to prevent the escape of the component parts of the plunger unit. Grasp the assembly through the cloth and pull it clear of the upper and lower lugs.

5   Examine the plunger guide rods for wear and damage. If worn, the rod must be renewed. Check the fit of the plunger on the rod. It should be a light sliding fit without appreciable free play. If the rods are rusted, due to lack of lubrication, it is permissible to remove this with fine emery cloth, provided that this does not result in the fit between the rod and plunger becoming sloppy.

6   Check the free length of the compression and rebound springs. When new, these springs measure $4\frac{7}{8}$ in (12.4 cms) and $2\frac{3}{4}$ (7 cms) respectively. If markedly shorter than this, or if they appear weakened or distorted, they should be renewed.

7   Before reassembling the units, check that the shrouds and gaiters are in good condition, paying particular attention to the latter. Temporarily refit the guide rod up through the lower lug, and assemble the plunger components over the rod in the order shown in the accompanying line drawing. When all the parts are in place, push down on the top of the unit until the assembly can be positioned between the two lugs.

8   Remove the guide rod, and refit it through the top frame lug. Screw the rod home, then refit and tighten the locknut. Fit the upper lug clamp bolt, and tighten it securely. Remember to lubricate each unit with a few strokes from a grease gun. This ensures free movement and prevents rusting of the guide rod.

14 Swinging arm rear suspension: examination and renovation

1   Later models of the Tiger Cub employed swinging arm suspension in which the wheel is carried in a tubular fork pivoted from the frame, the fork being supported by Girling hydraulically damped coil spring suspension units. The fork is bushed, and runs on a hardened steel pin pressed into the fr

lug. After an extended period of service, the bushes and the pin will wear, giving rise to lateral play in the fork assembly, and necessitating the renewal of the bushes, and possibly the pin. If not attended to when first noticed, the increasing amount of play will render the machine unsafe and cause it to fail the DOT test.

2   It is theoretically possible to renew the swinging arm bushes after driving the pivot pin out with a soft drift. In practice, the pin is often so firmly embedded in the lug, that a flypress or hydraulic press is necessary to effect its removal. It is not, therefore, recommended that the owner should tackle this job, as it is easy to cause irreparable damage when attempting to drive out a stubborn pin.

3   In the case of the machine used in the photographic sequences, removal and replacement of the pin and bushes proved impossible, using normal workshop tools. Most good motorcycle dealers will have the necessary facilities to complete this job economically. If the operation is attempted at home, it should be noted that the bushes will have to be line-reamed after fitting. At the time of writing, new bushes and pivot pins were still obtainable in the UK.

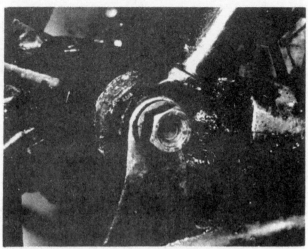

14.2 Remove bolt to allow frame members to be displaced

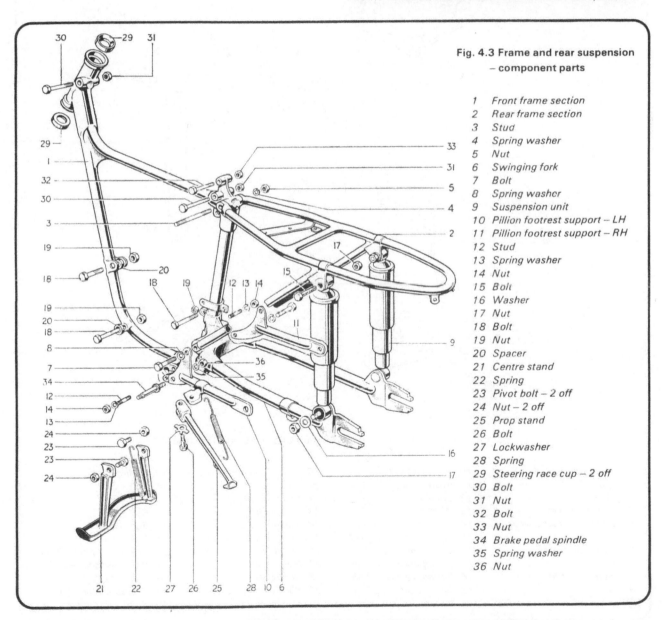

**Fig. 4.3 Frame and rear suspension
  – component parts**

1   Front frame section
2   Rear frame section
3   Stud
4   Spring washer
5   Nut
6   Swinging fork
7   Bolt
8   Spring washer
9   Suspension unit
10 Pillion footrest support – LH
11 Pillion footrest support – RH
12 Stud
13 Spring washer
14 Nut
15 Bolt
16 Washer
17 Nut
18 Bolt
19 Nut
20 Spacer
21 Centre stand
22 Spring
23 Pivot bolt – 2 off
24 Nut – 2 off
25 Prop stand
26 Bolt
27 Lockwasher
28 Spring
29 Steering race cup – 2 off
30 Bolt
31 Nut
32 Bolt
33 Nut
34 Brake pedal spindle
35 Spring washer
36 Nut

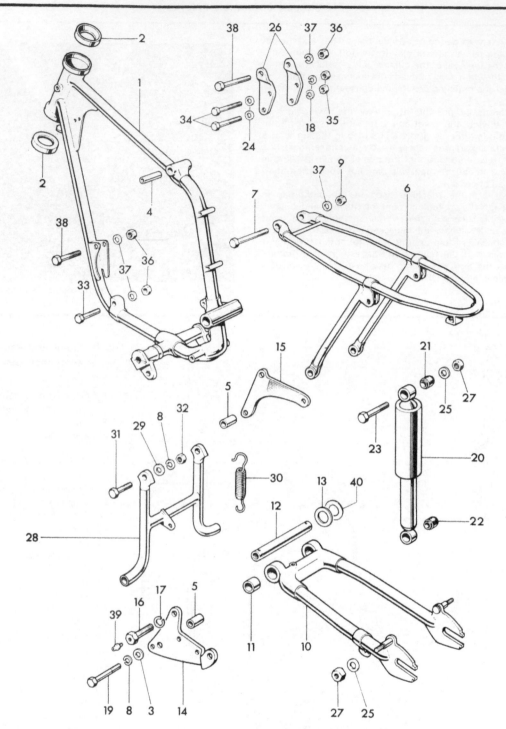

Fig. 4.4 Frame – component parts (Bantam Cub and Super Cub models)

1 Front frame assembly	10 Swinging fork	20 Suspension unit – 2 off	30 Return spring
2 Cup – 2 off	11 Bush – 2 off	21 Bush – 2 off	31 Bolt – 2 off
3 Washer	12 Spindle	22 Bush – 2 off	32 Nut – 2 off
4 Hollow dowel	13 Washer	23 Bolt – 2 off	33 Bolt
5 Pillion footrest spacer – 2 off	14 Left footrest bracket	24 Washer – 2 off	34 Bolt – 2 off
6 Rear frame assembly	15 Right footrest bracket	25 Plain washer – 4 off	35 Nut – 2 off
7 Bolt	16 Hollow bolt – 2 off	26 Engine rear plate – 2 off	36 Self-locking nut – 3 off
8 Spring washer – 5 off	17 Spring washer – 2 off	27 Nut – 4 off	37 Spring washer – 3 off
9 Nut	18 Spring washer – 2 off	28 Centre stand	38 Bolt – 2 off
	19 Bolt – 2 off	29 Plain washer – 2 off	39 Grease nipple – 2 off
			40 Shim – as required

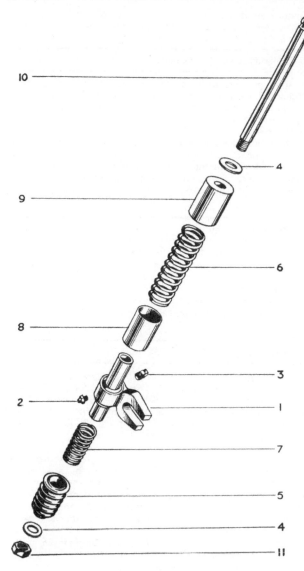

Fig. 4.5 Plunger rear suspension unit

1	Fork end – RH	6	Main spring
	Fork end – LH	7	Rebound spring
2	Grease nipple	8	Lower spring cover
3	Brake anchorage pin	9	Upper spring cover
4	Washer	10	Plunger rod
5	Plunger rubber gaiter	11	Locknut

15 Rear suspension units: examination

1   The Girling rear suspension units can be expected to give long service, and require no maintenance. In the event of failure, the machine will show a tendency to wallow when cornered, perhaps coupled with signs of oil leakage from the units.

2   No form of repair can be successfully performed at home, and the best course of action is to renew the units as a pair. In the event that replacement units of the original type are unobtainable, it may be possible to obtain a later type of unit which will fit. Before doing so, it is advisable to seek the advice of the manufacturer of the unit, to determine whether it will be suitable for the intended application.

3   It may be of interest to owners of competition machines that various proprietary suspension units are available for off-

15.1 Suspension units are each retained by bolt (top) and stud

road work. Again, it is wise to check the suitability of the proposed units prior to purchase.

16 Centre stand: examination

1   All models are provided with a centre stand attached to lugs on the bottom frame tubes. The stand provides a convenient means of parking the machine on level ground, or for raising one or other of the wheels clear of the ground in the event of a puncture. The stand pivots on a long bolt that passes through the lugs and is secured by a nut and washer. A return spring retracts the stand so that when the machine is pushed forward it will spring up and permit the machine to be wheeled, prior to riding.

2   The condition of the return spring and the return action should be checked frequently, also the security of the nut and bolt. If the stand drops whilst the machine is in motion, it may catch in some obstacle in the road and unseat the rider.

3   After many years of use, the centre stand stops and the corresponding lugs on the frame will wear, allowing the stand to go further over centre than is intended. If left to develop, gravity will eventually prevail and the owner will one day find his machine on its side. It is possible to have the stand stop and frame lugs made good by welding and filing to shape.

17 Prop stand: examination

1   A prop stand that pivots from a lug at the front end of the lower left-hand frame tube provides an additional means of parking the machine. This too has a return spring, which should be strong enough to cause the stand to retract immediately the machine is raised into a vertical position. It is important that this spring is examined at regular intervals, also the nut and bolt that act as the pivot. A falling prop stand can have more serious consequences if it should fall whilst the machine is on the move.

18 Footrests: examination and renovation

1   The footrests, which bolt to the frame lugs, are malleable and will bend if the machine is dropped. Before they can be straightened, they must be detached from the frame and have the rubbers removed.

2   To straighten the footrests, clamp them in a vice and apply leverage from a long tube that slips over the end. The area in which the bend has occurred should be heated to a cherry red with a blow lamp, during the bending operation. Do not bend the footrests cold, otherwise there is a risk of a sudden fracture.

19 Speedometer: removal and replacement

1   All models are equipped with a Smiths speedometer, mounted either in the headlamp nacelle or in a pressed steel casing attached to the upper fork yoke. The instrument head incorporates an odometer, which records the distance covered, and is fitted with an internal lamp which illuminates the dial when the lights are in use.

2   The speedometer is retained by two studs and a mounting strap, and may be released after the latter has been removed. On machines fitted with a nacelle, access is gained after removing the headlamp assembly. The drive cable is attached by a knurled ring on the underside of the case.

3   Apart from defects in the drive or the drive cable itself, a speedometer that malfunctions is difficult to repair. Fit a replacement or alternatively entrust the repair to an instrument repair specialist, bearing in mind that the speedometer must function in a satisfactory manner to meet Statutory requirements.

4   If the odometer readings continue to show an increase, without the speedometer indicating the road speed, it can be assumed the drive and drive cable are working correctly and that the speedometer head itself is at fault.

20 Speedometer cable: examination and renovation

1   It is advisable to detach the speedometer drive cable from time to time in order to check whether it is adequately lubricated, and whether the outer covering is compressed or damaged at any point along its run. A jerky or sluggish speedometer movement can often be attributed to a cable fault.

2   The cable is secured to the speedometer by a knurled ring, and runs the length of the machine to the drive gearbox at the rear wheel, to which it is attached by a hexagon-headed sleeve nut. The cable should be routed so that it does not pass through any sharp bends, which will cause accelerated wear.

21 Tachometer: removal and replacement

1   Some competition models are fitted with a tachometer, as well as, or instead of a speedometer, depending on the model. Drive is taken from a mechanism in the timing case, running off the unused distributor drive. The tachometer functions in a similar manner to the speedometer, measuring engine rather than road speed. Generally speaking, comments in Section 19 may be applied to the tachometer.

22 Tachometer drive cable: examination and renovation

1   Although a little shorter in length, the tachometer drive cable is identical in construction to that used for the speedometer drive. The advice given in Section 20 of this Chapter applies also to the tachometer drive cable.

23 Tank badges and knee pads

1   Although varying somewhat in design, the metal tank badges will be found to be retained by two screws on all models. It is unlikely that the badges will need to be removed, other than if the petrol tank requires refinishing, or if the badges themselves require renewal.

2   The rubber knee pads are retained by a moulded lip on the underside of each pad which engages in metal tangs welded to the tank sides. Note that if the tank is to be resprayed in its original colours, the paint beneath the knee pads will be relatively unfaded, and can be used for colour matching.

24 Steering head lock

1   Some models are fitted with a steering head lock inserted into the fork top yoke. If the forks are turned to the extreme left, they can be locked in this position to prevent theft.

2   Add an occasional few drops of thin machine oil to keep the lock in good working order. This should be added to the periphery of the moving drum and NOT the keyhole.

25 Cleaning: general

1   After removing all surface dirt with a rag or sponge that is washed frequently in clean water, the application of car polish or wax will restore a good finish to the cycle parts of the machine after they have dried thoroughly. The plated parts should require only a wipe with a damp rag although it is permissible to use a chrome cleaner if the plated surfaces are badly tarnished.

2   Oil and grease, particularly when they are caked on, are best removed with a proprietary cleanser such as 'Gunk' or 'Jizer'. A few minutes should be allowed for the cleanser to penetrate the film of oil and grease before the parts concerned are hosed down. Take care to protect the distributor, carburettor and electrical parts from the water, which may otherwise cause them to malfunction.

3   Polished aluminium alloy surfaces can be restored by the application of Solvol 'Autosol' or some similar polishing compound, and the use of a clean duster to give the final polish.

4   If possible, the machine should be wiped over immediately after it has been used in the wet, so that it is not garaged under damp conditions that will promote rusting. Make sure to wipe the chain and if necessary re-oil it, to prevent water from entering the rollers and causing harshness with an accompanying high rate of wear. Remember there is little chance of water entering the control cables if they are lubricated regularly, as recommended in the Routine Maintenance Section.

26 Fault diagnosis: frame and forks

Symptom	Cause	Remedy
Machine is unduly sensitive to road conditions	Forks and/or rear suspension units have defective damping	Check oil level in forks. Renew rear suspension units, if hydraulic type.
Machine tends to roll at low speeds	Steering head bearings overtight or damaged	Slacken bearing adjustment. If no improvement, dismantle and inspect bearings.
Machine tends to wander, steering is imprecise	Worn swinging arm bearings or sliders in plunger sprung models	Check and if necessary renew bearings.
Fork action stiff	Fork legs have twisted in yokes or have been drawn together at lower ends	Slacken off spindle nut clamps, pinch bolts in fork yokes and fork top nuts. Pump forks several times before retightening from bottom.
Forks judder when front brake is applied	Worn fork bushes Steering head bearings too slack	Strip forks and renew bushes. Readjust, to take up play.
Wheels out of alignment	Frame distorted as result of accident damage	Check frame alignment after stripping out. If bent, specialist repair is necessary.

Chapter 5 Wheels, brakes and tyres

Contents

General description 1
Front wheel: examination and renovation 2
Front wheel: removal and replacement 3
Front brake assembly: examination, renovation and
reassembly 4
Wheel bearings: examination and replacement 5
Rear wheel: examination, removal and replacement 6
Rear brake assembly: examination and renovation 7
Rear wheel bearings: removal, examination and
replacement 8

Front and rear brakes: adjustment 9
Rear wheel sprocket: removal, examination and
renovation 10
Final drive chain: examination, lubrication and adjustment 11
Wheel balance 12
Tyres: removal and replacement 13
Tyre valve dust caps 14
Fault diagnosis: wheels, brakes and tyres 15

Specifications

Tyre sizes

	Front	Rear
T.15	2.75 x 19 in	2.75 x 19 in
T.20 to 17388	3.00 x 19 in	3.00 x 19 in
T.20 from 17389 to 56360 ...	3.25 x 16 in	3.25 x 16 in
T.20 from 56361	3.25 x 17 in	3.25 x 17 in
T.20C	3.00 x 19 in	3.50 x 18 in
T.20T	3.00 x 19 in	3.50 x 18 in
T.20S	3.00 x 19 in	3.50 x 18 in
T.20SS	3.00 x 19 in	3.50 x 18 in
T.20SH	3.00 x 19 in	3.50 x 18 in

Tyre pressures

	Solo	Pillion
*Front (all models)	16 psi (1.13 kg cm^2)	20 psi (1.41 kg cm^2)
*Rear 3.50 x 18	16 psi (1.13 kg cm^2)	22 psi (1.55 kg cm^2)
All other sizes	18 psi (1.27 kg cm^2)	24 psi (1.70 kg cm^2)

Can be varied on trials models according to use

Brakes
Single leading shoe (sls) drum front and rear

1 General description

1 It will be noted that the rim and tyre sizes vary widely, according to the year of manufacture and model. Various diameters and sections were used, depending on the application to which the machine was put, and its overall weight. Note also that the tyre pressures quoted in the specifications can only be applied to road models, and even these may vary with different types of tyre. If in doubt, check the correct pressure with the tyre manufacturer or dealer. On competition machines, tyre pressure settings will have to be set according to the conditions of the event, and to some extent, to personal preference.

2 All models are equipped with a single leading shoe drum brake on each wheel, the two brakes being very similar and components such as shoes, interchangeable. Late models employ full-width aluminium alloy hubs, the actual working parts of the brake being very similar to the earlier half-width type. Some differences lie in the design of brake plates, due to the difference in the type of front fork used.

2 Front wheel: examination and renovation

1 Place the machine on the centre stand so that the front wheel is raised clear of the ground. Spin the wheel and check for rim alignment. Small irregularities can be corrected by tightening the spokes in the affected areas, although a certain amount of experience is necessary if over-correction is to be avoided. Any 'flats' in the wheel rim should be evident at the same time. These are more difficult to remove with any success and in most cases the wheel will have to be rebuilt on a new rim. Apart from the effect on stability, especially at high speeds, there is much greater risk of damage to the tyre beads and walls if the machine is ridden with a deformed wheel.

2 Check for loose or broken spokes. Tapping the spokes is the best guide to tension. A loose spoke will produce a quite different note and should be tightened by turning the nipple in an anti-clockwise direction. Always re-check for run-out by spinning the wheel again.

3 Front wheel: removal and replacement

1 Place the machine on the centre stand, if necessary placing blocks under the crankcase to raise the wheel clear of the ground. Disconnect the front brake cable at the actuating lever by pulling out the split pin and displacing the clevis pin. Unscrew the cable adjuster, and disengage the cable, if required.

2 On machines fitted with heavyweight forks, release the torque arm retaining bolt (lightweight forks have a lug which engages in a slot in the brake plate). Slacken and remove the four clamp nuts on the bottom of the fork legs, and lift the clamps away. The wheel can now be lowered and lifted clear of the forks.

3 Reassembly is a straightforward reversal of the removal sequence, bearing in mind the following points; ensure that the torque arm is positioned correctly, and that the bolt is refitted, where applicable. Tighten the clamp nuts evenly, so that the gap is even on each side of the spindle. Reconnect and adjust the brake cable before using the machine on the road.

4 Front brake assembly: examination, renovation and reassembly

1 The front wheel can be removed and the brake assembly complete with brake plate detached by following the procedure given in the preceding Section.

2 Before dismantling the brake assembly, examine the condition of the brake linings. If they are wearing thin or unevenly, they must be replaced.

3 To remove the brake shoes from the brake plate, position the operating cam so that the shoes are in the fully-expanded position and pull them apart whilst lifting them upwards, in the form of a 'V'. When they are clear of the brake plate the return springs should be removed and the shoes separated.

4 It is possible to renew the brake linings fitted with rivets and not bonded on, as is the current practice. Much will depend on the availability of the original type of linings; service-exchange brake shoes with bonded-on linings may be the only practical form of replacement.

5 If new linings are fitted by riveting, it is important that the rivet heads are countersunk, otherwise they will rub on the brake drum and be dangerous. Make sure the lining is in very close contact with the brake shoe during the riveting operation; a small 'C' clamp of the type used by carpenters can often be used to good effect until all the rivets are in position. Finish off by chamfering off the end of each shoe, otherwise fierce brake grab may occur due to pick-up of the leading edge of each lining.

6 Before replacing the brake shoes, check that the brake operating cam is working smoothly and not binding in its pivot.

The cam can be removed for greasing by unscrewing the nut on the end of the brake operating arm and drawing the arm off so that the cam and spindle can be withdrawn from the inside of the brake plate.

7 Check the inner surface of the brake drum, on which the brake shoes bear. The surface should be smooth and free from indentations, or reduced braking efficiency is inevitable. Remove all traces of brake lining dust and wipe the surface with a rag soaked in petrol to remove any traces of grease or oil.

8 To reassemble the brake shoes on the brake plate, fit the return springs and force the shoes apart, holding them in a 'V' formation. If they are now located with the brake operating arm and pivot, they can usually be snapped into position by pressing downward. Do not use excessive force, or the shoes may be distorted permanently.

5 Wheel bearings: examination and replacement

1 When the brake plate complete with brake assembly has been removed, the bearing retainer within the brake drum will be exposed. This ring has a **left-hand** thread, and should be unscrewed using a peg spanner where possible. Failing this, it is permissible to use a soft brass drift, noting that the ring itself is fairly soft and easily damaged. The bearing on the brake drum side of the wheel can now be displaced by driving the wheel spindle inwards. The spindle is shouldered, and will force the bearing out. The left-hand bearing is retained by a circlip, and may be drifted out of the hub using a drift, after the right-hand bearing has been removed. Note that the spindle is stepped and threaded at one end, and therefore neither it nor the bearings can be reversed or interchanged.

2 Remove all the old grease from the hub and bearings, giving the latter a final wash in petrol. Check the bearings for play or signs of roughness when they are turned. If there is any doubt about their condition, play safe and renew them. A new bearing has no discernible play.

3 Before replacing the bearings, first pack the hub with new, high melting point grease. Then grease both bearings and drive them back into position, not forgetting any distance piece, hollow sleeve or shim washers that were fitted originally. Make sure the bearing retainer is tight and that the dust cover is located correctly. The bearing retainer performs the dual role of preventing grease from entering the brake drum, thereby reducing braking efficiency.

4 There is no provision for adjusting wheel bearings of the journal ball type, therefore any discernible play indicates that the bearings have reached the end of their useful service life, and that renewal is necessary.

3.1 Release front brake cable from actuating lever and cable stop

3.2a Pin in backplate engages in slot on lightweight fork

3.2b Release clamps to allow wheel to be lowered

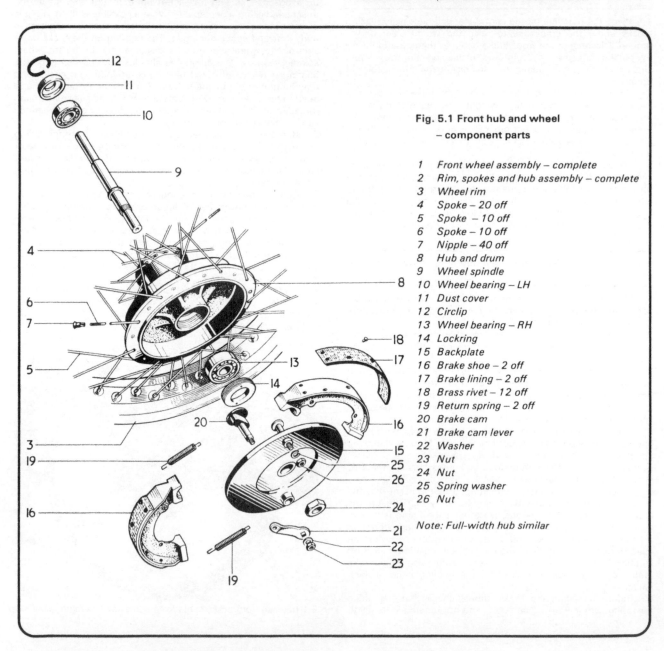

Fig. 5.1 Front hub and wheel
  – component parts

1   Front wheel assembly – complete
2   Rim, spokes and hub assembly – complete
3   Wheel rim
4   Spoke – 20 off
5   Spoke – 10 off
6   Spoke – 10 off
7   Nipple – 40 off
8   Hub and drum
9   Wheel spindle
10 Wheel bearing – LH
11 Dust cover
12 Circlip
13 Wheel bearing – RH
14 Lockring
15 Backplate
16 Brake shoe – 2 off
17 Brake lining – 2 off
18 Brass rivet – 12 off
19 Return spring – 2 off
20 Brake cam
21 Brake cam lever
22 Washer
23 Nut
24 Nut
25 Spring washer
26 Nut

Note: Full-width hub similar

6 Rear wheel: examination, removal and replacement

1 Place the machine on its centre stand, arranging it so that the rear wheel is raised clear of the ground. Check the wheel for rim alignment, damage and loose or broken spokes, adopting the same procedure as that described for front wheel examination in Section 2.

2 Separate the final drive chain at the joining link and disengage the chain from the rear wheel sprocket. Remove the torque arm mounting bolt. Slacken the drawbolt adjuster nuts. Remove the adjuster nut from the end of the brake operating rod, and allow the rod to disengage from the trunnion. It is a good idea to reassemble the brake rod components on the end of the rod to prevent their subsequent loss.

3 The two wheel spindle nuts can now be slackened to allow the wheel to be pulled back and disengaged from the fork ends. Note that it will be necessary either to remove the speedometer gearbox, or to release the speedometer drive cable, before the wheel is lifted away. The wheel is refitted by reversing the

6.2a Disconnect drive chain at joining link

6.2b Unscrew knurled adjuster to release brake rod

6.3a Wheel spindle has nut at each end ...

6.3b ... which must be released to allow wheel to be pulled clear

6.3c Tangs in speedometer drive gearbox engage in wheel

dismantling sequence, noting that chain adjustment, wheel alignment and brake adjustment should be carried out before the machine is used.

7   Rear brake assembly: examination and renovation

1   The brake plate assembly can be removed for examination after the wheel has been detached as described in the preceding Section. As mentioned previously, the front and rear brakes are virtually identical, and the comments given in Section 4 of this Chapter may be applied to the rear brake assembly.

8   Rear wheel bearings: removal, examination and replacement

1   The rear wheel bearings are similar in arrangement to those of the front hub, and may be dealt with in a similar manner, provided that the following points are noted; the speedometer drive gearbox must be removed, noting the two identical spacers which are fitted on either side of the unit. The bearing retainer ring (brake drum side) has a **right-hand** thread, and no circlip is used to retain the right-hand bearing. The bearings can be removed in the same way as that described for the front wheel bearings. Note that a plain washer is fitted behind the left-hand bearing.

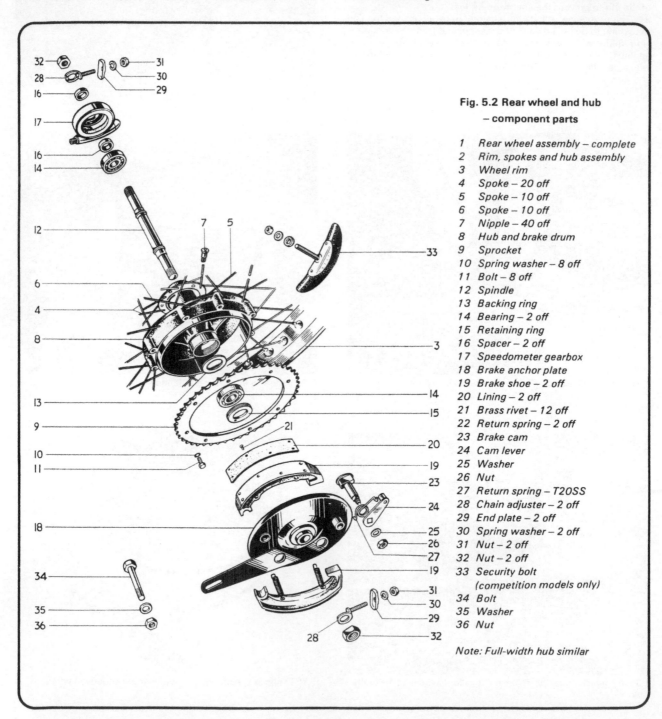

**Fig. 5.2 Rear wheel and hub
– component parts**

1   Rear wheel assembly – complete
2   Rim, spokes and hub assembly
3   Wheel rim
4   Spoke – 20 off
5   Spoke – 10 off
6   Spoke – 10 off
7   Nipple – 40 off
8   Hub and brake drum
9   Sprocket
10 Spring washer – 8 off
11 Bolt – 8 off
12 Spindle
13 Backing ring
14 Bearing – 2 off
15 Retaining ring
16 Spacer – 2 off
17 Speedometer gearbox
18 Brake anchor plate
19 Brake shoe – 2 off
20 Lining – 2 off
21 Brass rivet – 12 off
22 Return spring – 2 off
23 Brake cam
24 Cam lever
25 Washer
26 Nut
27 Return spring – T20SS
28 Chain adjuster – 2 off
29 End plate – 2 off
30 Spring washer – 2 off
31 Nut – 2 off
32 Nut – 2 off
33 Security bolt
     (competition models only)
34 Bolt
35 Washer
36 Nut

Note: Full-width hub similar

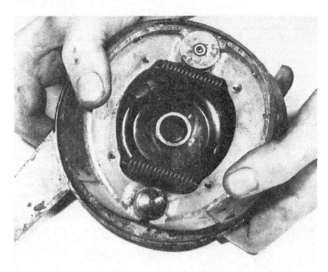

7.1a Brake plate assembly can be lifted out for examination

7.1b Fold shoes inwards to release them from brake plate

8.1a Note spacer fitted beneath speedometer drive gearbox

8.1b Rear wheel bearing retaining ring has RH thread

8.1c Clean bearings, then check for wear by spinning

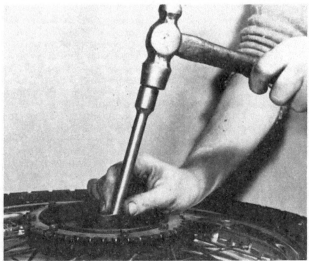

8.1d Worn bearings can be driven out of hub as shown

9 Front and rear brakes: adjustment

1   The setting of the brake lever and pedal is to some extent discretionary, but it is considered preferable to set each brake so that it commences operation as soon as the control is operated, as even a slight lag in brake operation can cause braking distances to increase appreciably. The front brake should be adjusted to give about $\frac{1}{16}$ - $\frac{1}{8}$ in free play in the cable.

2   The front brake cable adjuster is located at the lower end of the cable, and should be unscrewed until the brake just starts to rub. Turn the adjuster back until the required amount of slack is obtained. Rear brake adjustment is by way of a nut on the end of the operating rod. Set the rear brake to give about 1 in free movement at the pedal end.

3   No specific adjustment intervals can be given, as the need for adjustment is entirely dependent on the conditions the machine is used under, and the individual style of the rider. An experienced rider will quickly notice if brake adjustment becomes necessary, and will be able to correct this accordingly.

10 Rear wheel sprocket: removal, examination and renovation

1   The sprocket is attached to the rear wheel hub by eight bolts and spring washers, and is easily detached after the rear wheel has been released. The sprocket teeth should be checked carefully, and the component renewed if they appear hooked, chipped or badly worn, as a worn sprocket will quickly ruin the chain and the gearbox sprocket.

2   When obtaining a new replacement part, take the worn item as a pattern, or make a note of the number of teeth. Various combinations of sprocket sizes have been used over the years, and any departure from the standard arrangement will affect the overall gearing. Refer to Chapter 1 Specifications for details on chains and sprockets.

11 Final drive chain: examination, lubrication and adjustment

1   Unlike the primary chain, the final drive chain does not enjoy the benefits of full enclosure and an oil bath. In consequence, it will require periodic attention, particularly when the machine is used under inclement conditions.

2   Chain adjustment is correct when there is approximately $\frac{3}{4}$ inch play in the middle of the run. Always check at the tightest spot on the chain run, under load.

3   If the chain is too slack, adjustment is effected by slackening the rear wheel spindle and the torque arm, then drawing the wheel backwards by means of the chain adjusters at the end of the rear fork. Make sure each adjuster is turned an equal amount, so that the rear wheel is kept centrally-disposed within the frame. When the correct adjusting point has been reached, push the wheel hard forward to take up any slack, then tighten the spindle, not forgetting the torque arm nut. Re-check the chain tension and the wheel alignment, before the final tightening of the spindle and nuts. Wheel alignment can be checked by laying a straight wooden plank alongside the wheels, each side in turn – see accompanying illustrations.

4   Application of engine oil from time to time will serve as a satisfactory form of lubrication, but it is advisable to remove the chain every 2000 miles and clean it in a bath of paraffin before immersing it in a special chain lubricant such as 'Linklyfe' or 'Chainguard'. These latter types of lubricant achieve better and more lasting penetration of the chain links and rollers and are less likely to be thrown off when the chain is in motion. Intermediate lubrication should be carried out using one of the proprietary chain lubricants, which will penetrate and cling to the chain better than engine oil.

5   To check whether the chain is due for replacement, lay it lengthwise in a straight line and compress it, so that all play is taken up. Anchor one end and then pull on the other, to stretch the chain in the opposite direction. If the chain extends by more than $\frac{1}{4}$ inch per foot, replacement is necessary.

6   When replacing the chain, make sure the spring link is positioned correctly, with the closed end facing the direction of travel. Reconnection is made easier if the ends of the chain are pressed into the rear wheel sprocket.

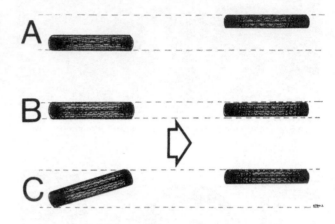

Fig. 5.3 Checking wheel alignment

A and C – Incorrect   *B – Correct*

12 Wheel balance

1   On any high performance machine it is important that the front wheel is balanced, to offset the weight of the tyre valve. If this precaution is not observed, the out-of-balance wheel will produce an unpleasant hammering that is felt through the handlebars at speeds from approximately 50 mph upwards.

2   To balance the front wheel, place the machine on the centre stand so that the front wheel is well clear of the ground and check that it will revolve quite freely, without the brake shoes rubbing. In the unbalanced state, it will be found that the wheel always comes to rest in the same position, with the tyre valve in the six o'clock position. Add balance weights to the spokes diametrically opposite the tyre valve until the tyre valve is counterbalanced, then spin the wheel to check that it will come to rest in a random position on each occasion. Add or subtract weight until perfect balance is achieved.

3   Only the front wheel requires attention. There is little point in balancing the rear wheel (unless both wheels are completely interchangeable) because it will have little noticeable effect on the handling of the machine.

4   Balance weights of various sizes that will fit around the spoke nipples were originally available from most motorcycle dealers. If difficulty is experienced in obtaining them, lead wire or even strip solder can be used as an alternative, kept in place with insulating tape.

13 Tyres: removal and replacement

1   At some time or other the need will arise to remove and replace the tyres, either as the result of a puncture or because replacements are necessary to offset wear. To the inexperienced, tyre changing represents a formidable task, yet if a few simple rules are observed and the technique learned, the whole operation is surprisingly simple.

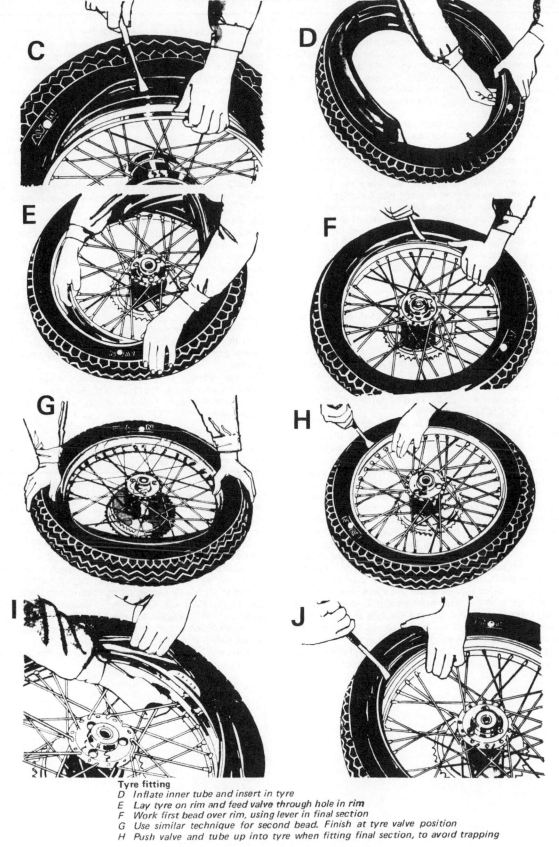

Tyre fitting
D Inflate inner tube and insert in tyre
E Lay tyre on rim and feed valve through hole in rim
F Work first bead over rim, using lever in final section
G Use similar technique for second bead. Finish at tyre valve position
H Push valve and tube up into tyre when fitting final section, to avoid trapping

Security bolts
I Fit the security bolt very loosely when one bead of the tyre is fitted.
J Then fit tyre in normal way. Tighten bolt when tyre is properly seated.

2   To remove the tyre from either wheel, first detach the wheel from the machine by following the procedure in Chapter 5.3 or 5.5, depending on whether the front or the rear wheel is involved. Deflate the tyre by removing the valve insert and when it is fully deflated, push the bead from the tyre away from the wheel rim on both sides so that the bead enters the centre well of the rim. Remove the locking cap and push the tyre valve into the tyre itself.

3   Insert a tyre lever close to the valve and lever the edge of the tyre over the outside of the wheel rim. Very little force should be necessary; if resistance is encountered it is probably due to the fact that the tyre beads have not entered the well of the wheel rim all the way round the tyre.

4   Once the tyre has been edged over the wheel rim, it is easy to work around the wheel rim so that the tyre is completely free on one side. At this stage, the inner tube can be removed.

5   Working from the other side of the wheel, ease the other edge of the tyre over the outside of the wheel rim that is furthest away. Continue to work around the rim until the tyre is free completely from the rim.

6   If a puncture has necessitated the removal of the tyre, re-inflate the inner tube and immerse it in a bowl of water to trace the source of the leak. Mark its position and deflate the tube. Dry the tube and clean the area around the puncture with a petrol soaked rag. When the surface has dried, apply rubber solution and allow this to dry before removing the backing from the patch and applying the patch to the surface.

7   It is best to use a patch of the self-vulcanising type, which will form a very permanent repair. Note that it may be necessary to remove a protective covering from the top surface of the patch, after it has sealed in position. Inner tubes made from synthetic rubber may require a special type of patch and adhesive, if a satisfactory bond is to be achieved.

8   Before replacing the tyre, check the inside to make sure the agent that caused the puncture is not trapped. Check the outside of the tyre, particularly the tread area, to make sure nothing is trapped that may cause a further puncture.

9   If the inner tube has been patched on a number of past occasions, or if there is a tear or large hole, it is preferable to discard it and fit a replacement. Sudden deflation may cause an accident, particularly if it occurs with the front wheel.

10 To replace the tyre, inflate the inner tube sufficiently for it to assume a circular shape but only just. Then push it into the tyre so that it is enclosed completely. Lay the tyre on the wheel at an angle and insert the valve through the rim tape and the hole in the wheel rim. Attach the locking cap on the first few threads, sufficient to hold the valve captive in its correct location.

11 Starting at the point furthest from the valve, push the tyre bead over the edge of the wheel rim until it is located in the central well. Continue to work around the tyre in this fashion until the whole of one side of the tyre is on the rim. It may be necessary to use a tyre lever during the final stages.

12 Make sure there is no pull on the tyre valve and again commencing with the area furthest from the valve, ease the other

bead of the tyre over the edge of the rim. Finish with the area close to the valve, pushing the valve up into the tyre until the locking cap touches the rim. This will ensure the inner tube is not trapped when the last section of the bead is edged over the rim with a tyre lever.

13 Check that the inner tube is not trapped at any point. Re-inflate the inner tube, and check that the tyre is seating correctly around the wheel rim. There should be a thin rib moulded around the wall of the tyre on both sides, which should be equidistant from the wheel rim at all points. If the tyre is unevenly located on the rim, try bouncing the wheel when the tyre is at the recommended pressure. It is probable that one of the beads has not pulled clear of the centre well.

14 Always run the tyres at the recommended pressures and never under or over-inflate. The correct pressures are given in the Specifications Section.

15 Tyre replacement is aided by dusting the side walls, particularly in the vicinity of the beads, with a liberal coating of french chalk. Washing-up liquid can also be used to good effect, but this has the disadvantage of causing the inner surfaces of the wheel rim to rust.

16 Never replace the inner tube and tyre without the rim tape in position. If this precaution is overlooked there is good chance of the ends of the spoke nipples chafing the inner tube and causing a crop of punctures.

17 Never fit a tyre that has damaged tread or side walls. Apart from the legal aspects, there is a very great risk of blow-out, which can have serious consequences on any two-wheel vehicle.

18 Tyre valves rarely give trouble, but it is always advisable to check whether the valve itself is leaking before removing the tyre. Do not forget to fit the dust cap, which forms an effective second seal.

14 Tyre valve dust caps

1   Tyre valve dust caps are often left off when a tyre has been replaced, despite the fact that they serve an important two-fold function. Firstly they prevent dirt or other foreign matter from entering the valve and causing the valve to stick open when the tyre pump is next applied. Secondly, they form an effective second seal so that in the event of the tyre valve sticking, air will not be lost.

2   Isolated cases of sudden deflation at high speeds have been traced to the omission of the dust cap. Centrifugal force has tended to lift the tyre valve off its seating and because the dust cap is missing, there has been no second seal. Racing inner tubes contain provision for this happening because the valve inserts are fitted with stronger springs, but standard inner tubes do not, hence the need for the dust cap.

3   Note that when a dust cap is fitted for the first time, the wheel may have to be rebalanced.

15 Fault diagnosis: wheels, brakes and tyres

Symptoms	Cause	Remedy
Handlebars oscillate at low speeds	Buckle or flat in wheel rim	Check rim alignment by spinning. Correct by retensioning spokes or by having wheel rebuilt on new rim.
	Tyre not straight on rim	Check tyre alignment.
Machine lacks power and accelerates poorly	Brakes binding	Warm brake drums provide best evidence. Re-adjust brakes.
Brakes grab, even when applied gently	Ends of brake shoes not chamfered	Chamfer with file.
	Elliptical brake drum	Lightly skim in lathe (specialist attention required).
Brake pull-off sluggish	Brake cam binding in housing	Free and grease.
	Weak brake shoe springs	Renew, if springs not displaced.
Harsh transmission	Worn or badly adjusted chains	Renew or adjust, as necessary.
	Hooked or badly worn sprockets	Renew as a pair.
	Rear wheel sprocket nuts loose	Check and tighten.
Middle of tyre treads wear rapidly	Tyres over-inflated	Check and readjust pressures.
Edges of tyre treads wear rapidly	Tyres under inflated	Check and increase pressures.
Forks hammer at high speeds	Front wheel not balanced	Balance wheel by adding balance weights.

Chapter 6 Electrical system

Contents

General description 1
Checking the charging system (Lucas type) 2
Rectifier: testing 3
Alternator and rectifier: location, removal and replacement 4
Lucas RM13 alternator system: alternative wiring 5
Battery: charging procedure and maintenance 6
Headlamp: replacing bulbs and adjusting beam height ... 7
Tail and stop lamp: replacing bulbs 8
Speedometer bulb: replacement 9
Horn: adjustment 10
Wiring: layout and examination 11
Headlamp switch 12
Ignition switch 13
Fault diagnosis: electrical system 14

Specifications

Battery
Make	Lucas
Type	PUZ5E/11
Voltage	6
Earth	Positive (+) earth

Alternator
Make	Lucas (Wipac on early models)
Type	RM 13
Output	6 volts ac

Rectifier
Make	Lucas
Type	47111A

Bulbs

	UK	USA	Europe
Headlamp	30/24 watt	30/24 watt	35/35 watt
Pilot	3 watt	–	3 watt
Speedometer	3 watt (where fitted	3 watt	3 watt
Rear lamp only	3 watt	3 watt	3 watt
Stop/tail lamp	6/18 watt	6/18 watt	6/18 watt

all bulbs rated 6 volt

1 General description

On all models, a six volt alternator mounted on the left hand side of the crankshaft provides power for the electrical and ignition systems. The ac current is fed to the rectifier, which converts it to dc current to be fed to the battery for charging purposes, and to the electrical system in general.

Output is controlled by switching in combinations of the stator coils, thus varying the amount of current available, to balance out the demands of the lighting system.

In the event of a discharged battery, an emergency ignition setting is provided by which the machine may be started directly from the alternator. On certain competition machines having an ET (energy transfer) ignition system, provision may be made for a simple lighting system running directly off the alternator.

2 Checking the charging system (Lucas type)

1 If charging system faults are suspected, it is essential to establish whether or not the battery is in good condition. In many cases, the alternator is blamed for problems caused by an ageing or sulphated battery. The easiest way to check this is to take the battery to an auto-electrician, who should be able to check the condition with a load tester and a hydrometer.

2 Check that all the electrical system connections are sound, and that there are no partial or complete shorts to earth caused by chafed wires. If the above measures fail to cure the problem, a full investigation of the system should be conducted. It is possible to do this at home, given the use of a test meter capable of measuring ac or dc volts and amps, such as a multimeter. A one ohm resistance will also be required. If in any doubt as to the method of operation of this equipment, it is

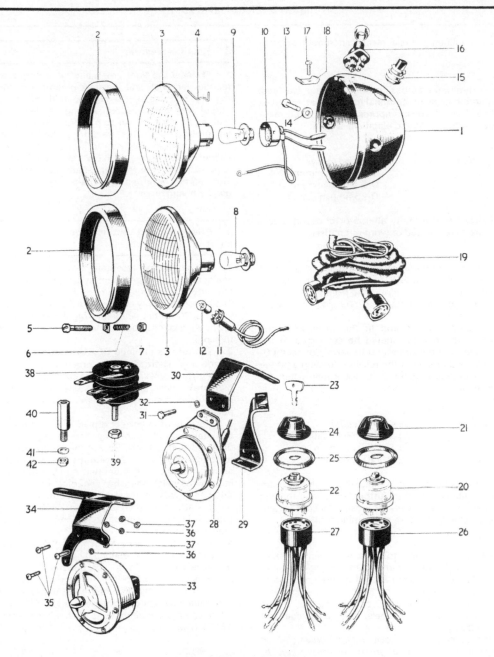

Fig. 6.1 Electrical equipment

1	Headlamp shell
2	Headlamp rim
3	Lens and reflector unit
4	W clip (as required)
5	Fixing and adjusting screw
6	Spring
7	Nut
8	Bulb 30/24w
9	Bulb 24/24w
10	Bulbholder
11	Pilot bulbholder
12	Pilot bulb 3w
13	Bolt (2 off)
14	Washer (2 off)
15	Lighting switch
16	Dip switch (T20SS, SH)
17	Screw
18	Retaining plate
19	Wiring harness
20	Lighting switch
21	Switch knob
22	Ignition switch
23	Ignition key
24	Switch knob
25	Bezel
26	Switch socket
27	Switch socket
28	Horn
29	Horn bracket
30	Horn bracket (alternative type)
31	Bolt (2 off)
32	Serrated washer
33	Horn (alternative type)
34	Horn bracket (alternative type)
35	Mounting screw (3 off)
36	Serrated washer (3 off)
37	Nut (3 off)
38	Rectifier
39	Nut
40	Extended nut
41	Serrated washer
42	Nut

suggested that a qualified vehicle electrican should be consulted.

Test A

3   Connect a dc ammeter across the battery negative feed, after disconnecting the negative terminal from the battery. Start the engine and run it at 3000 rpm approx. The following readings may be expected from a sound electrical system:

Lighting switch position	Reading (Amps)
OFF	1·5 amp minimum
PILOT	0·5 amp minimum
HEAD	0·25 amp minimum

No reading is indicative of a faulty alternator or circuit, a low reading can indicate a discharged or worn out battery.

Test B

4   Disconnect the battery negative lead. Connect a length of wire between the ignition coil SW terminal, and the battery negative terminal. In this way, current is fed from the battery to the ignition system, but the lighting and charging circuits are isolated from the battery.
5   Obtain a one ohm resistor, and fit this in place of the battery. One end of the resistor should be connected to the battery negative lead, and the other end to earth. Connect a 0-10 volt dc voltmeter either side of the resistor, forming a bridge across it. Start the engine and run it at 3000 rpm, when the various readings should be as follows:

Lighting switch position	Reading (volts)
OFF	1.5 volts minimum
PILOT	1.5 volts minimum
HEAD	3.0 volts minimum

6   If a low reading is obtained, it is indicative of an alternator defect, and this should be checked further. An erratic reading can often be traced to chafed wires or loose connections, and these areas must be checked before proceeding further.

Test C

7   The alternator will be found to have three output leads, these being coloured; light green, dark green and mid greeen or green and yellow. The output leads should be disconnected at the bullet connectors, and tested for no-load voltage as follows:
8   Obtain a 0-10 volt ac voltmeter, and connect a 1 ohm resistor in parallel across its terminals. The four stage test should be connected as follows:
1   *Probe leads connected to: dark green and light green*
2   *Probe leads connected to: light green and mid green (or green and yellow)*
3   *Probe leads connected to: dark green and light green with mid green (or green and yellow) connected to dark green*
4   *One probe lead to any one alternator output lead, the other probe lead to earth*
    Conduct each stage separately, noting the reading obtained at 3000 rpm approx. Readings should be as follows:

Test stage	Reading (volts ac)
1	3.0 volts minimum
2	6.0 volts minimum
3	8.5 volts minimum
4	No reading

9   A low reading from stage 1, 2 and 3 indicates earthed or shorted stator windings, a zero reading indicating an open circuit or shorted coil. If all of the first three readings are low, it is possible that the rotor has become partially de-magnetised. This can be caused by a faulty rectifier or an incorrectly connected battery, and these areas must be checked before fitting a new rotor. Any reading other than zero at stage 4 indicates an earthed coil. Check and repair the earth leak or renew the stator.

3   Rectifier: testing

1   Connect the light green alternator output lead to the No 1 rectifier terminal, and the dark green alternator output lead to the No 3 terminal, leaving the mid green (or green/yellow) output lead disconnected. Arrange a 0–10 volt dc voltmeter with a one ohm resistor as a bridge across its terminal. Connect one probe to earth, and the other to the No 2 rectifier terminal.
2   Start the engine and run it at 3000 rpm when the reading should be a minimum of 6.5 volts. If this figure is not obtained, it indicates that the rectifier is faulty or is not effectively earthed. This can be confirmed by substituting a rectifier known to be in good working order.

4   Alternator and rectifier: location, removal and replacement

1   The alternator assembly is contained within the primary chaincase, the stator assembly being mounted on the inside of the outer casing, and the rotor being keyed to the crankshaft. The stator assembly can be detached after releasing the mounting nuts. Note that when refitting the assembly, ensure that the connecting leads between the stator coils face inwards, towards the engine.
2   The rotor is retained by a Woodruff key to the tapered end of the mainshaft. The securing nut can be slackened after locking the crankshaft by selecting top gear and applying the rear brake. Where more than one keyway is used, mark the relevant one before drawing the rotor off with a three-legged puller.
3   Note that the correct replacement rotor must be used with the stator fitted, and if necessary advice on this should be obtained from a Lucas Service Agent. When refitting the rotor, the Lucas motif should face outwards. For reference, the stator assembly is supplied under part number 465915 (47138 on ET ignition models), the corresponding rotor part number being 466124. Lucas alternators are to some extent interchangeable, but it is advisable to check the suitability of a proposed alternative, as the wiring connections and/or associated components may have to be changed.
4   The rectifier is located beneath the dualseat (or saddle, on early vehicles) and is retained by a central mounting bolt.
Important note: On no account must the nut which secures and tensions the rectifier plates be disturbed, or the unit will be ruined. The plate tension is pre-set during manufacture, and is not adjustable. On early models, a No 47103 rectifier may be fitted, and if this malfunctions, it should be replaced by the later 47111 type.

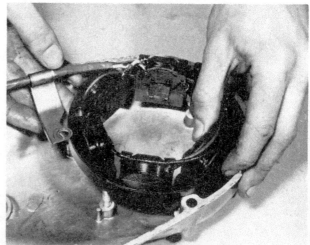

4.1 Alternator stator is retained by studs and nuts

4.2 Rotor is keyed to crankshaft taper

5 Lucas RM 13 alternator system: alternative wiring

1   On early machines fitted with the RM 13 system, a bridge link was incorporated in the lighting switch between terminals 5 and 6. This link was omitted on later machines, to give a higher charge rate with the lights off. If still fitted, this link should be removed to give increased output.

2   Where the machine is used for short journeys with frequent use of lights, the alternator should be connected to give maximum output. The appropriate connections are as follows:

Alternator output cable to	Wiring harness cable
mid green (or green/yellow)	dark green
dark green	mid green
light green	light green

3   Where much long distance work, without lights is contemplated, the leads should be connected colour to colour to prevent over-charging. This condition is indicated by excessive gassing in the battery, coupled with the need for frequent topping-up.

4   The engine may be run for short periods with the battery removed, provided that the battery negative lead is connected to earth. Note that no lighting will be available in this mode, and that if lights are fitted to the machine, they must be in working order, in the UK and many other countries. Prolonged use of the machine wired in this way can cause rectifier damage, and so should be avoided.

5   It is possible to fit a later 12 volt alternator system if extra power is felt necessary. Bear in mind that the alternator, rectifier, battery horn, coil and all bulbs must be changed for 12 volt components, and this will involve a considerable outlay. Obtain advice from a Lucas Service Agent concerning the correct replacement parts for this conversion.

6 Battery: charging procedure and maintenance

1   Whilst the machine is used on the road it is unlikely that the battery will require attention other than routine maintenance because the generator will keep it fully charged. However, if the machine is used for a succession of short journeys only, mainly during the hours of darkness when the lights are in full use, it is possible that the output from the generator may fail to keep pace with the heavy electrical demand, especially if the machine is parked with the lights switched on. Under these circumstances, it will be necessary to remove the battery from time to time to have it charged independently. Note the comments on charge rates described in Section 5.

2   The battery is located inside the toolbox, in a carrier, It is secured by a strap which, when released, will permit the battery to be withdrawn after disconnection of the leads. The battery positive is always earthed. To remove the battery release the carrier clamp and disconnect the electrical leads. Some models may have a different form of battery holder, although the battery will be found in approximately the same location below the dual seat.

3   The normal charge rate is 1 amp. A more rapid charge can be given in an emergency, but this should be avoided if possible because it will shorten the life of the battery.

4   When the battery is removed from the machine, remove the cover and clean the battery top. If the terminals are corroded, scrape them clean and cover them with vaseline (not grease) to protect them from further attack. If a vent tube is fitted, make sure it is not obstructed and that it is arranged so that it will not discharge over any parts of the machine.

5   If the machine is laid up for any period of time, the battery should be removed and given a 'refresher' charge every six weeks or so, in order to maintain it in good condition.

6   Several different types of Lucas lead-acid battery have been fitted to the various models since their inception. A variety of replacement types is available at widely differing prices. Generally, quality is in direct proportion to the purchase price.

7   Battery maintenance is limited to keeping the electrolyte level just above the plates and separators. Modern batteries have a transparent plastic case with a level line, which makes the check of the electrolyte level much easier.

8   Unless acid is spilt, which may occur if the machine falls over, use only distilled water for topping up purposes, until the correct level is restored. If acid is spilt on any part of the machine, it should be neutralised immediately with an alkali such as washing soda or baking powder, and washed away with plenty of water. This will prevent corrosion from taking place. Top up in this instance with sulphuric acid of the correct specific gravity (1.260 - 1.280).

9   It is seldom practicable to repair a cracked battery case because the acid that is already seeping through the crack will prevent the formation of an effective seal, no matter what sealing compound is used. It is always best to replace a cracked battery especially in view of the risk of corrosion from the acid leakage.

10 Make sure that the battery is clamped securely. A loose battery will vibrate and its working life will be greatly shortened, due to the paste being shaken out of the plates.

7 Headlamp: replacing bulbs and adjusting beam height

1   Although the type of mounting varies between models, the lens and reflector unit is similar in all cases, and access to the bulbs is gained after releasing the unit from the nacelle or shell. On nacelle models, release the spring adjusting screw from the underside of the unit, and lift the assembly away. On models fitted with a conventional shell, release the rim securing screw and pull the unit away from the shell.

2   Release the bulb holder by pushing and turning anticlockwise. The prefocus bulb can then be withdrawn. Replacement and reassembly is a straightforward reversal of the removal sequence.

3   Beam height on nacelle models is changed by moving the spring adjuster screw on the underside of the unit. On machines with a separate shell, adjustment is accomplished by slackening the two bolts that retain the headlamp shell to the forks and tilting the headlamp either upwards or downwards. Adjustments should always be made with the rider seated normally.

4   UK lighting regulations stipulate that the lighting system must be arranged so that the light does not dazzle a person standing in the same horizontal plane as the vehicle, at a distance greater than 25 feet from the lamp. whose eye level is not less than 3 feet 6 inches above that plane. It is easy to approximate this setting by placing the machine 25 feet away

from a wall, on a level road and setting the beam height so that it is concentrated at the same height as the distance from the centre of the headlamp to the ground. The rider must be seated normally during this operation, and the pillion passenger, if one is carried regularly.

5   If the headlamp reflector unit is broken, it can be removed from the headlamp rim by detaching the wire retaining clips, after the front has been removed from the headlamp. In the case of a headlamp of the unit glass/reflector type, it will be necessary to purchase the complete beam unit, and not the glass alone.

8   Tail and stop lamp: replacing bulbs

1   Although the very early models were supplied with a tail lamp only, it is doubtful whether any of the original fittings are still in use because the size no longer meets the minimum requirements of the lighting regulations. Most of the current tail lamp units contain provision for a stop lamp bulb also, which is operated when the rear brake pedal is depressed.

2   Removal of the plastic lens cover will reveal the bulb holders for both the tail lamp and the stop lamp, which may be separate, as in the case of the early models or combined, to conform to current practice. It is now customary to fit a single bulb with offset pins, which is of the double filament type. The offset pins prevent accidental inversion of the bulb. The tail lamp filament is rated at 6W and the stop lamp filament at 18W.

3   The stop lamp switch will be found on the left hand side of the machine, in close proximity to the brake pedal. The switch does not require attention other than the occasional drop of light oil.

9   Speedometer bulb: replacement

1   The bulb that illuminates the dial of this instrument has a bayonet fitting in a metal bulbholder that pushes into the base of the instrument case. On nacelle models, the speedometer is illuminated by reflected light from the headlamp unit.

10 Horn: adjustment

1   A horn of the electro-magnetic type is fitted to every machine. It is operated from a push button mounted on the handlebars.

2   A small, serrated screw, located in the back of the horn body, provides means of adjusting the note. It must not be turned more than two or three notches, before re-testing. If the ammeter shows a reading when the horn is operated it should not exceed 4 amps. It is possible to obtain a good note, at the expense of an excessive current requirement.

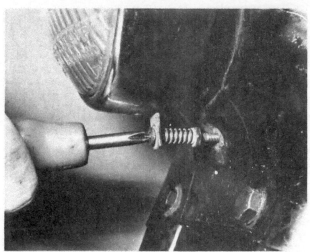

7.1 Spring-loaded screw retains headlamp and provides adjustment

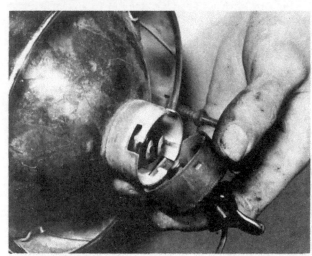

7.2a Bulb is retained by bayonet cap

7.2b Pilot bulbholder is a push-fit in the headlamp reflector

8.2 Rear lamp lens is secured by two nuts
may be used

11 Wiring: layout and examination

1   The cables of the wiring harness are colour-coded and will correspond with the accompanying wiring diagrams.

2   Visual inspection will show whether any break or frayed outer coverings are giving rise to short circuits which will cause the main fuse to blow. Another source of trouble is the snap connectors and spade terminals, which may make a poor connection if they are not pushed home fully.

3   Intermittent short circuits can sometimes be traced to a chafed wire passing through, or close by, to a metal component, such as a frame member. Avoid tight bends in the cables or situations where the cables can be trapped or stretched, especially in the vicinity of the handlebars or steering head.

12 Headlamp switch

1   It is unusual for the headlamp switch to give trouble unless the machine has been laid up for a considerable period and the switch contacts have become dirty. Contact between the terminal posts is made by a spring-loaded roller attached to the body of the switch knob. If the terminal posts become corroded or oxidised, poor or intermittent electrical contact will result.

2   It is possible to dismantle the switch and clean the terminal posts and roller by hand; the rotor containing the roller can be pulled away when the centre screw of the switch knob is removed and the knob detached. This is a somewhat delicate operation, which should be performed only when the switch complete is removed from the headlamp shell. The switch body is held to the underside of the shell by means of a wire spring that engages with a groove around the body moulding.

3   A better alternative that does not necessitate dismantling the switch is the use of one of the proprietary switch contact cleansers that are available in aerosol form.

4   On no account oil the switch or oil will spread across the internal contacts, to form an effective insulator.

13 Ignition switch

1   Where a separate ignition switch is fitted, it is similar in construction to the headlamp switch. The remarks in Section 12 may be applied to its maintenance and repair.

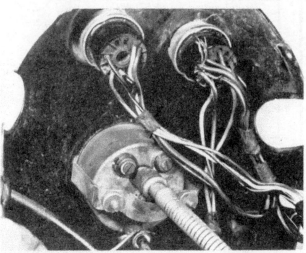

12.2 Separate lighting and ignition switches, or a combined unit,

14 Fault diagnosis: electrical system

Symptom	Cause	Remedy
Complete electrical failure	Short circuit	Check wiring and electrical components for insulation breakdown.
	Isolated battery	Check battery connections, also whether connections show signs of corrosion.
Dim lights, horn inoperative	Discharged battery	Recharge battery with battery charger and check whether alternator is giving correct output.
Constantly 'blowing' bulbs	Vibration, poor earth connection	Check whether bulb holders are secured correctly. Check earth return or connections to frame.

RED

TAIL LAMP
6 VOLT 3 WATT

BROWN/GREEN

RED

HEADLAMP
6 VOLT
24/24 WATT

STOP LAMP
6 VOLT 18 WATT

BROWN

BROWN/BLUE

ON/OFF
SWITCH

DIPPER
SWITCH

BLUE/WHITE

BLUE/WHITE

STOPLIGHT
SWITCH

SINGLE SNAP
CONNECTOR

BROWN

HORN

BROWN/
BLACK

SPEEDOMETER
LAMP

BROWN/GREEN

BROWN/
BLUE

BLUE
IDENT

BROWN/
BLUE

3 WAY SNAP
CONNECTOR

BROWN/BLACK

HORN PUSH

SINGLE SNAP
CONNECTORS

BROWN

DOUBLE SNAP
CONNECTOR

BLACK/WHITE

CUT-OUT
BUTTON

BROWN

BROWN/
BLUE

DOUBLE SNAP
CONNECTOR

BLACK/WHITE

L.T.

ALTERNATOR

TO SPARK
PLUG

SINGLE SNAP
CONNECTOR

COIL

H.T.

CONTACT
BREAKER

E

RED

RED

Wiring diagram – machines with AC ignition and direct lighting

STOP LAMP SWITCH

STOP LAMP

BROWN/
BLUE

BROWN/
BLUE

BROWN

BROWN

BROWN

HORN PUSH

BROWN/
BLACK

LIGHTING SWITCH

GREEN/WHITE

HORN

ALTERNATOR

BLACK

5 4 3

2

6

RED/BLACK

BROWN/BLUE

6 VOLT BATTERY

7

1

GREEN/
BLACK

GREEN/YELLOW

WHITE/GREEN

BROWN/
GREEN

8 10 11

BROWN/GREEN

BROWN/BLUE

RED

GREEN/
BLACK

BLUE

HEADLAMP MAIN BEAM

WHITE/GREEN

BLUE/
WHITE

BLUE/
WHITE

RED

BROWN/
BLUE

BLUE

HEADLAMP DIP SWITCH

GREEN/
BLACK

BROWN/
BLUE

DIPPER
SWITCH

BLUE/RED

BLUE/
RED

RED

PILOT LIGHT

RED/BLACK

RED/BLACK

BROWN/
BLUE

12 12 19

BROWN/
BLUE

BLACK

SPEEDOMETER LAMP

13

BROWN/GREEN

13A

18

IGNITION
SWITCH

TAIL LAMP

14

17

BROWN/GREEN

BROWN/GREEN

BROWN/GREEN

15 16

RED

BLACK/WHITE

(SW) –

COIL

RECTIFIER

(CB) +

SNAP CONNECTORS

RED

EARTH CONNECTIONS
MADE VIA CABLE
OR
VIA FIXING BOLTS

WHITE

BLACK/
WHITE

CONTACT
BREAKER

Wiring diagram – machines equipped with separate lighting and ignition switches

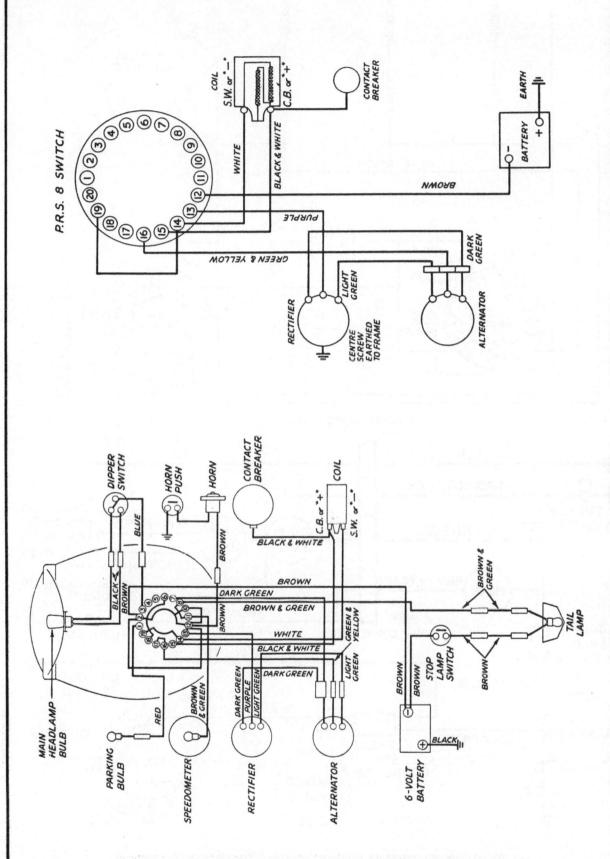

Optional wiring diagram – machines using battery/coil ignition without lights

Wiring diagram – machines equipped with combined ignition/lighting switch

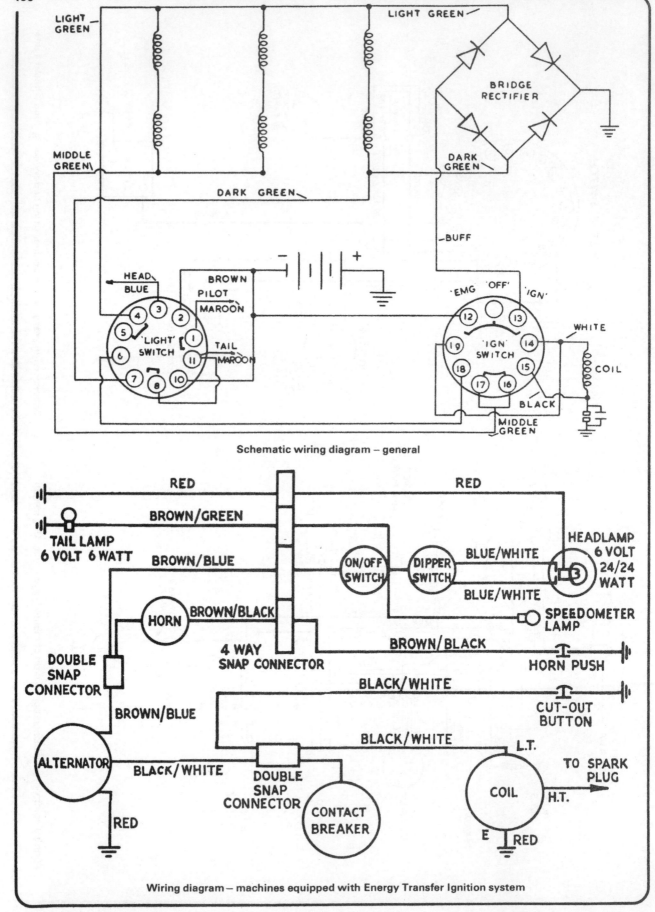

Schematic wiring diagram – general

Wiring diagram – machines equipped with Energy Transfer Ignition system

Conversion factors

Length (distance)

Inches (in)	X	25.4	= Millimetres (mm)	X	0.0394	= Inches (in)
Feet (ft)	X	0.305	= Metres (m)	X	3.281	= Feet (ft)
Miles	X	1.609	= Kilometres (km)	X	0.621	= Miles

Volume (capacity)

Cubic inches (cu in; in³)	X	16.387	= Cubic centimetres (cc; cm³)	X	0.061	= Cubic inches (cu in; in³)
Imperial pints (Imp pt)	X	0.568	= Litres (l)	X	1.76	= Imperial pints (Imp pt)
Imperial quarts (Imp qt)	X	1.137	= Litres (l)	X	0.88	= Imperial quarts (Imp qt)
Imperial quarts (Imp qt)	X	1.201	= US quarts (US qt)	X	0.833	= Imperial quarts (Imp qt)
US quarts (US qt)	X	0.946	= Litres (l)	X	1.057	= US quarts (US qt)
Imperial gallons (Imp gal)	X	4.546	= Litres (l)	X	0.22	= Imperial gallons (Imp gal)
Imperial gallons (Imp gal)	X	1.201	= US gallons (US gal)	X	0.833	= Imperial gallons (Imp gal)
US gallons (US gal)	X	3.785	= Litres (l)	X	0.264	= US gallons (US gal)

Mass (weight)

Ounces (oz)	X	28.35	= Grams (g)	X	0.035	= Ounces (oz)
Pounds (lb)	X	0.454	= Kilograms (kg)	X	2.205	= Pounds (lb)

Force

Ounces-force (ozf; oz)	X	0.278	= Newtons (N)	X	3.6	= Ounces-force (ozf; oz)
Pounds-force (lbf; lb)	X	4.448	= Newtons (N)	X	0.225	= Pounds-force (lbf; lb)
Newtons (N)	X	0.1	= Kilograms-force (kgf; kg)	X	9.81	= Newtons (N)

Pressure

Pounds-force per square inch (psi; lbf/in²; lb/in²)	X	0.070	= Kilograms-force per square centimetre (kgf/cm²; kg/cm²)	X	14.223	= Pounds-force per square inch (psi; lbf/in²; lb/in²)
Pounds-force per square inch (psi; lbf/in²; lb/in²)	X	0.068	= Atmospheres (atm)	X	14.696	= Pounds-force per square inch (psi; lbf/in²; lb/in²)
Pounds-force per square inch (psi; lbf/in²; lb/in²)	X	0.069	= Bars	X	14.5	= Pounds-force per square inch (psi; lbf/in²; lb/in²)
Pounds-force per square inch (psi; lbf/in²; lb/in²)	X	6.895	= Kilopascals (kPa)	X	0.145	= Pounds-force per square inch (psi; lbf/in²; lb/in²)
Kilopascals (kPa)	X	0.01	= Kilograms-force per square centimetre (kgf/cm²; kg/cm²)	X	98.1	= Kilopascals (kPa)

Torque (moment of force)

Pounds-force inches (lbf in; lb in)	X	1.152	= Kilograms-force centimetre (kgf cm; kg cm)	X	0.868	= Pounds-force inches (lbf in; lb in)
Pounds-force inches (lbf in; lb in)	X	0.113	= Newton metres (Nm)	X	8.85	= Pounds-force inches (lbf in; lb in)
Pounds-force inches (lbf in; lb in)	X	0.083	= Pounds-force feet (lbf ft; lb ft)	X	12	= Pounds-force inches (lbf in; lb in)
Pounds-force feet (lbf ft; lb ft)	X	0.138	= Kilograms-force metres (kgf m; kg m)	X	7.233	= Pounds-force feet (lbf ft; lb ft)
Pounds-force feet (lbf ft; lb ft)	X	1.356	= Newton metres (Nm)	X	0.738	= Pounds-force feet (lbf ft; lb ft)
Newton metres (Nm)	X	0.102	= Kilograms-force metres (kgf m; kg m)	X	9.804	= Newton metres (Nm)

Power

Horsepower (hp)	X	745.7	= Watts (W)	X	0.0013	= Horsepower (hp)

Velocity (speed)

Miles per hour (miles/hr; mph)	X	1.609	= Kilometres per hour (km/hr; kph)	X	0.621	= Miles per hour (miles/hr; mph)

Fuel consumption*

Miles per gallon, Imperial (mpg)	X	0.354	= Kilometres per litre (km/l)	X	2.825	= Miles per gallon, Imperial (mpg)
Miles per gallon, US (mpg)	X	0.425	= Kilometres per litre (km/l)	X	2.352	= Miles per gallon, US (mpg)

Temperature

Degrees Fahrenheit = (°C x 1.8) + 32   Degrees Celsius (Degrees Centigrade; °C) = (°F - 32) x 0.56

It is common practice to convert from miles per gallon (mpg) to litres/100 kilometres (l/100km), where mpg (Imperial) x l/100 km = 282 and mpg (US) x l/100 km = 235

English/American terminology

Because this book has been written in England, British English component names, phrases and spellings have been used throughout. American English usage is quite often different and whereas normally no confusion should occur, a list of equivalent terminology is given below.

English	American	English	American
Air filter	Air cleaner	Mudguard	Fender
Alignment (headlamp)	Aim	Number plate	License plate
Allen screw/key	Socket screw/wrench	Output or layshaft	Countershaft
Anticlockwise	Counterclockwise	Panniers	Side cases
Bottom/top gear	Low/high gear	Paraffin	Kerosene
Bottom/top yoke	Bottom/top triple clamp	Petrol	Gasoline
Bush	Bushing	Petrol/fuel tank	Gas tank
Carburettor	Carburetor	Pinking	Pinging
Catch	Latch	Rear suspension unit	Rear shock absorber
Circlip	Snap ring	Rocker cover	Valve cover
Clutch drum	Clutch housing	Selector	Shifter
Dip switch	Dimmer switch	Self-locking pliers	Vise-grips
Disulphide	Disulfide	Side or parking lamp	Parking or auxiliary light
Dynamo	DC generator	Side or prop stand	Kick stand
Earth	Ground	Silencer	Muffler
End float	End play	Spanner	Wrench
Engineer's blue	Machinist's dye	Split pin	Cotter pin
Exhaust pipe	Header	Stanchion	Tube
Fault diagnosis	Trouble shooting	Sulphuric	Sulfuric
Float chamber	Float bowl	Sump	Oil pan
Footrest	Footpeg	Swinging arm	Swingarm
Fuel/petrol tap	Petcock	Tab washer	Lock washer
Gaiter	Boot	Top box	Trunk
Gearbox	Transmission	Two/four stroke	Two/four cycle
Gearchange	Shift	Tyre	Tire
Gudgeon pin	Wrist/piston pin	Valve collar	Valve retainer
Indicator	Turn signal	Valve collets	Valve cotters
Inlet	Intake	Vice	Vise
Input shaft or mainshaft	Mainshaft	Wheel spindle	Axle
Kickstart	Kickstarter	White spirit	Stoddard solvent
Lower leg	Slider	Windscreen	Windshield

Index

A

Acknowledgements 2
About this manual 2
Adjustments:-
    brakes 12, 94
    carburettor 63
    clutch 12, 50
    contact breaker 11, 69
    final drive chain 94
    final engine 50
    headlamp beam height 101
    horn 102
    ignition timing 71
    rear chain tension 10
    valve clearances 11, 16
    valve rocker clearance 50
Air filter element 63
Alternator 100

B

Battery charging procedure 101
Balancing front wheel 35
Bearings:-
    big end 30
    main 30
    steering head 81
    wheel:-
        front 89
        rear 92
Brakes:-
    adjusting 12, 94
    front – examination, renovation and reassembly 89
    rear – examination and renovation 92
    pedal – rear brake 9, 12
Bulbs:-
    replacement:-
        headlamp 101
        speedometer 102
        stop and tail 102
Buying:-
    spare parts 8
    tools 14

C

Carburettors:-
    adjustment 58
    dismantling and reassembly 55
    fault diagnosis 66
    removal 55
    settings 63
    specifications 54

Cables:-
    brake 76
    clutch 12, 50
    lubrication 12
    speedometer and tachometer 86
    throttle 56
Chapter contents:-
    1 Engine, clutch and gearbox 15
    2 Fuel system and lubrication 54
    3 Ignition system 68
    4 Frame and forks 75
    5 Wheels, brakes and tyres 88
    6 Electrical system 98
Checks:-
    Brake linings 12
    carburettor settings 63
    charging system Lucas type 98
    coil 73
    contact breaker points 11, 69
    engine – oil level 9
    general 50
    ignition timing 11, 71
    legal 9
    lubricating system 63
    safety 9
    settings – carburettor 63
    sparking plug – gap setting 74
    tyre pressures 9, 88
    valve clearance 11, 16
    wheel bearings 12
    wheel spokes 90
Centre stand 85
Cleaning:-
    air filter element 63
    carburettor 12
    rear chain 10
    sparking plug 11
    the machine 86
    wheel bearings 12
Clutch:-
    adjustment 11
    cable 12
    drag 53
    examination and renovation 38
    fault diagnosis 53
    refitting 46
    slip 38
    specification 17
Chain – rear drive 10
Coil – ignition 73
Condenser 71

Connecting rod 19, 33
Contact breaker 69
Crankcase halves:-
  separating 30
  joining 40
Crankshaft 17
Cylinder barrel 19, 34, 48
Cylinder head 19, 34, 44

D

Decarbonising 34, 35
Description – general:-
  electrical system 98
  engine, clutch and gearbox 17
  frame and forks 75
  fuel system 55
  ignition system 68
  lubrication 55
  wheels, brakes and tyres 88
Dimensions and weights 6
Distributor 20, 71
Dust caps – tyre valve 96

E

Electrical system:-
  alternator 100
  battery 101
  fault diagnosis 103
  headlamp 101
  horn 102
  lamps 101–102
  rectifier 100
  specification 98
  switches – headlamp and ignition 113
  wiring diagrams 101–106
Engine:-
  alternator rotor 20
  camshaft, tappets and oil pump 24
  clutch – examination and renovation 38
  connecting rod 19, 33
  crankshaft 17
  crankcase halves:-
    separating 30
    joining 40
  crankcase types 25
  cylinder barrel 19, 34
  cylinder head 19, 34, 49
  decarbonising 34, 35
  dismantling – general 19
  distributor 20
  examination and renovation – general 30
  fault diagnosis 52
  gudgeon pin 19, 33, 34, 48
  oil pump:-
    removal 64
    fitting 41
  piston and rings 19, 34, 48
  pushrods 17, 19
  reassembly – general 40
  refitting in frame 49
  removal from frame 17
  small end bush 33
  specifications 15–16
  starting and running the rebuilt unit 50
  tappets 41
  valve grinding 34, 35

F

Fault diagnosis:-
  clutch 53
  electrical system 103
  engine 52
  frame and forks 87
  fuel system 66
  gearbox 53
  ignition system 74
  lubrication system 66
  wheels, brakes and tyres 97
Filters:-
  air 63
  oil 55
Frame and forks:-
  centre stand 85
  fault diagnosis 87
  footrests 85
  frame examination and renovation 82
  front forks – dismantling 78
  oil seals – front forks 78
  prop stand 85
  rear brake pedal 12
  rear suspension unit 85
  speedometer and tachometer drive cables 86
  steering head bearings 81
  steering head lock 86
  swinging arm suspension 82
Front wheel 89
Fuel system:-
  air filter element 63
  carburettor:-
    adjustment 63
    dismantling and examination 55–56
    settings – checking 63
    fault diagnosis 66
    petrol feed pipes 55
    petrol tank 55
    petrol tap 55

G

Gearbox:-
  components removal 22
  fault diagnosis 53
  refitting components 12
  specifications 16
Generator – alternator 100
Gudgeon pin 19, 33, 34, 48

H

Headlamp beam height adjustment 101
Horn 102

I

Ignition system:-
  A.T.U. 69
  coil – checking 73
  condenser 71
  contact breaker adjustment 69
  distributor 71
  fault diagnosis 74
  sparking plug:-
    checking and setting gap 11, 74
    operating conditions:-
      colour chart 67
  specification 68
  switch 103
  timing 11, 71

K

Kickstart 36, 42

L

Legal obligation 9, 96, 101
Lighting and ignition switch 103
Lubrication system 55

Lubrication:-
    cables 12
    general 10
    rear chain 10
Lubrication – recommended 13

M

Maintenance – routine 9, 12
Modifications to the Triumph Tiger and Terrier Cub
    Range 7

O

Oil filter gauze 55
Oil pump 24, 41
Ordering:-
    spare parts 8

P

Pedal – rear brake 12, 94
Petrol feed pipes 55
Petrol tank 55
Petrol tap 55
Pistons and rings 19, 34, 38
Prop stand 85

R

Rear brake pedal 12, 95
Rear chain – cleaning 10
Rear suspension unit 85
Rear wheel 91
Rear wheel sprocket 94
Rectifier 100
Recommended lubricants 13
Rings and pistons 19, 34, 38
Routine maintenance 9–12

S

Safety check 9
Sparking plug:-
    gap settings 11, 74
    operating conditions – colour chart 67
Speedometer 86

Specifications:-
    bulbs 98
    clutch 17
    electrical system 98
    engine 15–16
    frame and forks 75
    fuel system 54
    gearbox 16
    ignition 68
    wheels, brakes and tyres 88
Statutory requirements 10, 96, 101
Steering head lock 86
Suspension unit – rear 85
Switches:-
    headlamps 103
    ignition 103

T

Tachometer 86
Timing ignition 11, 71
Tools 14
Tyres:-
    pressures 9, 88
    removal and replacement 94
    colour instructions 87
    valves and dust caps 96

V

Valves – engine
    clearance 11, 16
    grindings 34–35
    guides 34–35
    springs 34–35
Valves and dust caps 96

W

Weight and dimensions 6
Wheels:-
    balancing – front wheel 94
    bearings 89, 92
    front 89
    rear 91
    spokes 89
Wiring diagrams 104–106
Wiring layout and examination 103
Working conditions 14